www.EffortlessMath.com

... So Much More Online!

✓ FREE Math lessons

✓ More Math learning books!

✓ Mathematics Worksheets

✓ Online Math Tutors

Need a PDF version of this book?

Please visit www.EffortlessMath.com

ASVAB Math Prep 2020-2021

The Most Comprehensive Review and Ultimate Guide to the ASVAB Math Test

By

Reza Nazari & Ava Ross

Copyright © 2020

Reza Nazari & Ava Ross

All rights reserved. No part of this publication may be reproduced, stored in a retrieval system, or transmitted in any form or by any means, electronic, mechanical, photocopying, recording, scanning, or otherwise, except as permitted under Section 107 or 108 of the 1976 United States Copyright Ac, without permission of the author.

All inquiries should be addressed to:

info@effortlessMath.com

www.EffortlessMath.com

ISBN: 978-1-64612-300-1

Published by: Effortless Math Education

www.EffortlessMath.com

Visit www.EffortlessMath.com
for Online Math Practice

Description

ASVAB Math Prep 2020 – 2021, which reflects the 2020 - 2021 test guidelines, is dedicated to preparing test takers to ace the ASVAB Math Test. This comprehensive ASVAB Math Prep book with hundreds of examples, abundant sample ASVAB mathematics questions, and two full-length and realistic ASVAB Math tests is all you will ever need to fully prepare for the ASVAB Math. It will help you learn everything you need to ace the math section of the ASVAB test.

Effortless Math unique study program provides you with an in-depth focus on the math portion of the exam, helping you master the math skills that students find the most troublesome. This ASVAB Math preparation book contains most common sample questions that are most likely to appear in the mathematics section of the ASVAB.

Inside the pages of this comprehensive ASVAB Math book, students can learn basic math operations in a structured manner with a complete study program to help them understand essential math skills. It also has many exciting features, including:

- ✓ Content 100% aligned with the 2020 ASVAB test
- ✓ Written by ASVAB Math instructors and test experts
- ✓ Complete coverage of all ASVAB Math concepts and topics which you will be tested
- ✓ Over 2,500 additional ASVAB math practice questions in both multiple-choice and grid-in formats with answers grouped by topic, so you can focus on your weak areas
- ✓ Abundant Math skill building exercises to help test-takers approach different question types that might be unfamiliar to them
- ✓ Exercises on different ASVAB Math topics such as integers, percent, equations, polynomials, exponents and radicals
- ✓ 2 full-length practice tests (featuring new question types) with detailed answers

ASVAB Math Prep 2020 – 2021 is an incredibly useful resource for those who want to review all topics being covered on the ASVAB test. It efficiently and effectively reinforces learning outcomes through engaging questions and repeated practice, helping you to quickly master Math skills.

About the Author

Reza Nazari is the author of more than 100 Math learning books including:
– **Math and Critical Thinking Challenges:** For the Middle and High School Student
– **ACT Math in 30 Days**
– **ASVAB Math Workbook 2018 - 2019**
– **Effortless Math Education Workbooks**
– and many more Mathematics books …

Reza is also an experienced Math instructor and a test–prep expert who has been tutoring students since 2008. Reza is the founder of Effortless Math Education, a tutoring company that has helped many students raise their standardized test scores—and attend the colleges of their dreams. Reza provides an individualized custom learning plan and the personalized attention that makes a difference in how students view math.

You can contact Reza via email at:
reza@EffortlessMath.com

Find Reza's professional profile at:
goo.gl/zoC9rJ

Contents

Chapter 1: Fractions and Mixed Numbers 8

Chapter 2: Decimals 19

Chapter 3: Integers and Order of Operations 26

Chapter 4: Ratios and Proportions 33

Chapter 5: Percentage 39

Chapter 6: Expressions and Variables 47

Chapter 7: Equations and Inequalities 56

Chapter 8: Lines and Slope 66

Chapter 9: Exponents and Variables 76

Chapter 10: Polynomials 87

Chapter 11: Geometry and Solid Figures 98

Chapter 12: Statistics 111

Chapter 13: Functions Operations 119

ASVAB Test Review 126

ASVAB Math Practice Test 1 CAT-ASVAB Arithmetic Reasoning 130

ASVAB Math Practice Test 1 CAT-ASVAB Mathematics Knowledge 135

ASVAB Math Practice Test 2 Paper and Pencil-ASVAB Arithmetic Reasoning .. 139

ASVAB Math Practice Test 2 Paper and Pencil-ASVAB Mathematics Knowledge 147

ASVAB Mathematics Practice Tests Answers and Explanations 153

Chapter 1:

Fractions and Mixed Numbers

Math Topics that you'll learn in this Chapter:

- ✓ Simplifying Fractions
- ✓ Adding and Subtracting Fractions
- ✓ Multiplying and Dividing Fractions
- ✓ Adding Mixed Numbers
- ✓ Subtracting Mixed Numbers
- ✓ Multiplying Mixed Numbers
- ✓ Dividing Mixed Numbers

Simplifying Fractions

✓ A fraction contains two numbers separated by a bar in between them. The bottom number, called the denominator, is the total number of equally divided portions in one whole. The top number, called the numerator, is how many portions you have. And the bar represents the operation of division.

✓ Simplifying a fraction means reducing it to lowest terms. To simplify a fraction, evenly divide both the top and bottom of the fraction by 2, 3, 5, 7, ... etc.

✓ Continue until you can't go any further.

Examples:

1) Simplify $\frac{12}{30}$

   Solution: To simplify $\frac{12}{30}$, find a number that both 12 and 30 are divisible by.

   Both are divisible by 6. Then: $\frac{12}{30} = \frac{12 \div 6}{30 \div 6} = \frac{2}{5}$

2) Simplify $\frac{64}{80}$

   Solution: To simplify $\frac{64}{80}$, find a number that both 64 and 80 are divisible by.

   Both are divisible by 8 and 16. Then: $\frac{64}{80} = \frac{64 \div 8}{80 \div 8} = \frac{8}{10}$, 8 and 10 are divisible by 2,

   then: $\frac{8}{10} = \frac{4}{5}$ or $\frac{64}{80} = \frac{64 \div 16}{80 \div 16} = \frac{4}{5}$

3) Simplify $\frac{20}{60}$

   Solution: To simplify $\frac{20}{60}$, find a number that both 20 and 60 are divisible by. Both are divisible by 20.

   Then: $\frac{20}{60} = \frac{20 \div 20}{60 \div 20} = \frac{1}{3}$

Adding and Subtracting Fractions

- For "like" fractions (fractions with the same denominator), add or subtract the numerators (top numbers) and write the answer over the common denominator (bottom numbers).

- Adding and Subtracting fractions with the same denominator:

$$\frac{a}{b} + \frac{c}{b} = \frac{a+c}{b} \qquad \frac{a}{b} - \frac{c}{b} = \frac{a-c}{b}$$

- Find equivalent fractions with the same denominator before you can add or subtract fractions with different denominators.

- Adding and Subtracting fractions with different denominators:

$$\frac{a}{b} + \frac{c}{d} = \frac{ad+bc}{bd} \qquad \frac{a}{b} - \frac{c}{d} = \frac{ad-bc}{bd}$$

Examples:

1) Find the sum. $\frac{3}{4} + \frac{1}{3} =$

    Solution: These two fractions are "unlike" fractions. (they have different denominators). Use this formula: $\frac{a}{b} + \frac{c}{d} = \frac{ad+cb}{bd}$

    Then: $\frac{3}{4} + \frac{1}{3} = \frac{(3)(3)+(4)(1)}{4 \times 3} = \frac{9+4}{12} = \frac{13}{12}$

2) Find the difference. $\frac{4}{5} - \frac{3}{7} =$

    Solution: For "unlike" fractions, find equivalent fractions with the same denominator before you can add or subtract fractions with different denominators. Use this formula:

    $\frac{a}{b} - \frac{c}{d} = \frac{ad-bc}{bd}$

    $\frac{4}{5} - \frac{3}{7} = \frac{(4)(7)-(3)(5)}{5 \times 7} = \frac{28-15}{35} = \frac{13}{35}$

Multiplying and Dividing Fractions

☑ Multiplying fractions: multiply the top numbers and multiply the bottom numbers. Simplify if necessary. $\frac{a}{b} \times \frac{c}{d} = \frac{a \times c}{b \times d}$

☑ Dividing fractions: Keep, Change, Flip

Keep first fraction, change division sign to multiplication, and flip the numerator and denominator of the second fraction. Then, solve! $\frac{a}{b} \div \frac{c}{d} = \frac{a}{b} \times \frac{d}{c} = \frac{a \times d}{b \times c}$

Examples:

1) Multiply. $\frac{5}{8} \times \frac{2}{3} =$

   Solution: Multiply the top numbers and multiply the bottom numbers.
   $\frac{5}{8} \times \frac{2}{3} = \frac{5 \times 2}{8 \times 3} = \frac{10}{24}$, simplify: $\frac{10}{24} = \frac{10 \div 2}{24 \div 2} = \frac{5}{12}$

2) Solve. $\frac{1}{3} \div \frac{4}{7} =$

   Solution: Keep first fraction, change division sign to multiplication, and flip the numerator and denominator of the second fraction.
   Then: $\frac{1}{3} \div \frac{4}{7} = \frac{1}{3} \times \frac{7}{4} = \frac{1 \times 7}{3 \times 4} = \frac{7}{12}$

3) Calculate. $\frac{3}{5} \times \frac{2}{3} =$

   Solution: Multiply the top numbers and multiply the bottom numbers.
   $\frac{3}{5} \times \frac{2}{3} = \frac{3 \times 2}{5 \times 3} = \frac{6}{15}$, simplify: $\frac{6}{15} = \frac{6 \div 3}{15 \div 3} = \frac{2}{5}$

4) Solve. $\frac{1}{4} \div \frac{5}{6} =$

   Solution: Keep first fraction, change division sign to multiplication, and flip the numerator and denominator of the second fraction.
   Then: $\frac{1}{4} \div \frac{5}{6} = \frac{1}{4} \times \frac{6}{5} = \frac{1 \times 6}{4 \times 5} = \frac{6}{20}$, simplify: $\frac{6}{20} = \frac{6 \div 2}{20 \div 2} = \frac{3}{10}$

Adding Mixed Numbers

Use following steps for adding mixed numbers:

- Add whole numbers of the mixed numbers.
- Add the fractions of the mixed numbers.
- Find the Least Common Denominator (LCD) if necessary.
- Add whole numbers and fractions.
- Write your answer in lowest terms.

Examples:

1) Add mixed numbers. $3\frac{1}{3} + 1\frac{4}{5} =$

   Solution: Let's rewriting our equation with parts separated, $3\frac{1}{3} + 1\frac{4}{5} = 3 + \frac{1}{3} + 1 + \frac{4}{5}$. Now, add whole number parts: $3 + 1 = 4$

   Add the fraction parts $\frac{1}{3} + \frac{4}{5}$. Rewrite to solve with the equivalent fractions. $\frac{1}{3} + \frac{4}{5} = \frac{5}{15} + \frac{12}{15} = \frac{17}{15}$. The answer is an improper fraction (numerator is bigger than denominator). Convert the improper fraction into a mixed number: $\frac{17}{15} = 1\frac{2}{15}$. Now, combine the whole and fraction parts: $4 + 1\frac{2}{15} = 5\frac{2}{15}$

2) Find the sum. $1\frac{2}{5} + 2\frac{1}{2} =$

   Solution: Rewriting our equation with parts separated, $1 + \frac{2}{5} + 2 + \frac{1}{2}$. Add the whole number parts:

   $1 + 2 = 3$. Add the fraction parts: $\frac{2}{5} + \frac{1}{2} = \frac{4}{10} + \frac{5}{10} = \frac{9}{10}$

   Now, combine the whole and fraction parts: $3 + \frac{9}{10} = 3\frac{9}{10}$

Subtract Mixed Numbers

Use the following steps for subtracting mixed numbers.

✓ Convert mixed numbers into improper fractions. $a\frac{c}{b} = \frac{ab+c}{b}$

✓ Find equivalent fractions with the same denominator for unlike fractions. (fractions with different denominators)

✓ Subtract the second fraction from the first one. $\frac{a}{b} - \frac{c}{d} = \frac{ad-bc}{bd}$

✓ Write your answer in lowest terms.

✓ If the answer is an improper fraction, convert it into a mixed number.

Examples:

1) Subtract. $3\frac{4}{5} - 1\frac{3}{4} =$

   Solution: Convert mixed numbers into fractions: $3\frac{4}{5} = \frac{3\times5+4}{5} = \frac{19}{5}$ and $1\frac{3}{4} = \frac{1\times4+3}{4} = \frac{7}{4}$. These two fractions are "unlike" fractions. (they have different denominators). Find equivalent fractions with the same denominator. Use this formula: $\frac{a}{b} - \frac{c}{d} = \frac{ad-bc}{bd}$
   $\frac{19}{5} - \frac{7}{4} = \frac{(19)(4)-(5)(7)}{5\times4} = \frac{76-35}{20} = \frac{41}{20}$, the answer is an improper fraction, convert it into a mixed number. $\frac{41}{20} = 2\frac{1}{20}$

2) Subtract. $4\frac{3}{8} - 1\frac{1}{2} =$

   Solution: Convert mixed numbers into fractions: $4\frac{3}{8} = \frac{4\times8+3}{8} = \frac{35}{8}$ and $1\frac{1}{2} = \frac{1\times2+1}{4} = \frac{3}{2}$
   Find equivalent fractions: $\frac{3}{2} = \frac{12}{8}$. Then: $4\frac{3}{8} - 1\frac{1}{2} = \frac{35}{8} - \frac{12}{8} = \frac{23}{8}$
   The answer is an improper fraction, convert it into a mixed number.
   $$\frac{23}{8} = 2\frac{7}{8}$$

Multiplying Mixed Numbers

Use following steps for multiplying mixed numbers:

✓ Convert the mixed numbers into fractions. $a\frac{c}{b} = a + \frac{c}{b} = \frac{ab+c}{b}$

✓ Multiply fractions. $\frac{a}{b} \times \frac{c}{d} = \frac{a \times c}{b \times d}$

✓ Write your answer in lowest terms.

✓ If the answer is an improper fraction (numerator is bigger than denominator), convert it into a mixed number.

Examples:

1) Multiply. $3\frac{1}{3} \times 4\frac{1}{6} =$

   Solution: Convert mixed numbers into fractions, $3\frac{1}{3} = \frac{3 \times 3 + 1}{3} = \frac{10}{3}$ and $4\frac{1}{6} = \frac{4 \times 6 + 1}{6} = \frac{25}{6}$

   Apply the fractions rule for multiplication, $\frac{10}{3} \times \frac{25}{6} = \frac{10 \times 25}{3 \times 6} = \frac{250}{18}$

   The answer is an improper fraction. Convert it into a mixed number. $\frac{250}{18} = 13\frac{8}{9}$

2) Multiply. $2\frac{1}{2} \times 3\frac{2}{3} =$

   Solution: Converting mixed numbers into fractions, $2\frac{1}{2} \times 3\frac{2}{3} = \frac{5}{2} \times \frac{11}{3}$

   Apply the fractions rule for multiplication, $\frac{5}{2} \times \frac{11}{3} = \frac{5 \times 11}{2 \times 3} = \frac{55}{6} = 9\frac{1}{6}$

3) Multiply mixed numbers. $2\frac{1}{3} \times 2\frac{1}{2} =$

   Solution: Converting mixed numbers to fractions, $2\frac{1}{3} = \frac{7}{3}$ and $2\frac{1}{2} = \frac{5}{2}$. Multiply two fractions:

   $$\frac{7}{3} \times \frac{5}{2} = \frac{7 \times 5}{3 \times 2} = \frac{35}{6} = 5\frac{5}{6}$$

Dividing Mixed Numbers

Use following steps for dividing mixed numbers:

- Convert the mixed numbers into fractions. $a\frac{c}{b} = a + \frac{c}{b} = \frac{ab+c}{b}$

- Divide fractions: Keep, Change, Flip: Keep first fraction, change division sign to multiplication, and flip the numerator and denominator of the second fraction. Then, solve! $\frac{a}{b} \div \frac{c}{d} = \frac{a}{b} \times \frac{d}{c} = \frac{a \times d}{b \times c}$

- Write your answer in lowest terms.

- If the answer is an improper fraction (numerator is bigger than denominator), convert it into a mixed number.

Examples:

1) Solve. $3\frac{2}{3} \div 2\frac{1}{2}$

   Solution: Convert mixed numbers into fractions: $3\frac{2}{3} = \frac{3 \times 3 + 2}{3} = \frac{11}{3}$ and $2\frac{1}{2} = \frac{2 \times 2 + 1}{2} = \frac{5}{2}$

   Keep, Change, Flip: $\frac{11}{3} \div \frac{5}{2} = \frac{11}{3} \times \frac{2}{5} = \frac{11 \times 2}{3 \times 5} = \frac{22}{15}$. The answer is an improper fraction. Convert it into a mixed number: $\frac{22}{15} = 1\frac{7}{15}$

2) Solve. $3\frac{4}{5} \div 1\frac{5}{6}$

   Solution: Convert mixed numbers to fractions, then solve:

   $3\frac{4}{5} \div 1\frac{5}{6} = \frac{19}{5} \div \frac{11}{6} = \frac{19}{5} \times \frac{6}{11} = \frac{114}{55} = 2\frac{4}{55}$

3) Solve. $2\frac{2}{7} \div 2\frac{3}{5}$

   Solution: Converting mixed numbers to fractions: $3\frac{4}{5} \div 1\frac{5}{6} = \frac{16}{7} \div \frac{13}{5}$

   Keep, Change, Flip: $\frac{16}{7} \div \frac{13}{5} = \frac{16}{7} \times \frac{5}{13} = \frac{16 \times 5}{7 \times 13} = \frac{80}{91}$

Chapter 1: Practices

✎ *Simplify each fraction.*

1) $\dfrac{18}{30} =$

2) $\dfrac{21}{42} =$

3) $\dfrac{35}{55} =$

4) $\dfrac{48}{72} =$

5) $\dfrac{54}{81} =$

6) $\dfrac{80}{200} =$

✎ *Find the sum or difference.*

7) $\dfrac{6}{15} + \dfrac{3}{15} =$

8) $\dfrac{2}{3} + \dfrac{1}{9} =$

9) $\dfrac{1}{4} + \dfrac{2}{5} =$

10) $\dfrac{7}{10} - \dfrac{3}{10} =$

11) $\dfrac{1}{2} - \dfrac{3}{8} =$

12) $\dfrac{5}{7} - \dfrac{3}{5} =$

✎ *Find the answers.*

13) $\dfrac{1}{7} \div \dfrac{3}{8} =$

14) $\dfrac{2}{3} \times \dfrac{4}{7} =$

15) $\dfrac{5}{7} \times \dfrac{3}{4} =$

16) $\dfrac{2}{5} \div \dfrac{3}{7} =$

17) $\dfrac{3}{7} \div \dfrac{5}{8} =$

18) $\dfrac{3}{8} \times \dfrac{4}{7} =$

✎ *Calculate.*

19) $3\dfrac{1}{5} + 2\dfrac{2}{9} =$

20) $1\dfrac{1}{7} + 5\dfrac{2}{5} =$

21) $4\dfrac{4}{5} + 1\dfrac{2}{7} =$

22) $2\dfrac{4}{7} + 2\dfrac{3}{5} =$

23) $1\dfrac{5}{6} + 1\dfrac{2}{5} =$

24) $3\dfrac{5}{7} + 1\dfrac{2}{9} =$

✎ *Calculate.*

25) $3\frac{2}{5} - 1\frac{2}{9} =$

27) $4\frac{2}{5} - 2\frac{2}{7} =$

29) $9\frac{5}{7} - 7\frac{4}{21} =$

26) $5\frac{3}{5} - 1\frac{1}{7} =$

28) $8\frac{3}{4} - 2\frac{1}{8} =$

30) $11\frac{7}{12} - 9\frac{5}{6} =$

✎ *Find the answers.*

31) $1\frac{1}{8} \times 1\frac{3}{4} =$

33) $2\frac{1}{8} \times 1\frac{2}{9} =$

35) $1\frac{1}{2} \times 5\frac{2}{3} =$

32) $3\frac{1}{5} \times 2\frac{2}{7} =$

34) $2\frac{3}{8} \times 2\frac{2}{5} =$

36) $3\frac{1}{2} \times 6\frac{2}{3} =$

✎ *Solve.*

37) $9\frac{1}{2} \div 2\frac{3}{5} =$

39) $5\frac{3}{4} \div 2\frac{2}{7} =$

41) $7\frac{2}{5} \div 3\frac{3}{4} =$

38) $2\frac{3}{8} \div 1\frac{2}{5} =$

40) $8\frac{1}{3} \div 4\frac{1}{4} =$

42) $2\frac{4}{5} \div 3\frac{2}{3} =$

Answers – Chapter 1

1) $\frac{3}{5}$
2) $\frac{1}{2}$
3) $\frac{7}{11}$
4) $\frac{2}{3}$
5) $\frac{2}{3}$
6) $\frac{2}{5}$
7) $\frac{3}{5}$
8) $\frac{7}{9}$
9) $\frac{13}{20}$
10) $\frac{2}{5}$
11) $\frac{1}{8}$
12) $\frac{4}{35}$
13) $\frac{8}{21}$
14) $\frac{8}{21}$
15) $\frac{15}{28}$
16) $\frac{14}{15}$
17) $\frac{24}{35}$
18) $\frac{3}{14}$
19) $5\frac{19}{45}$
20) $6\frac{19}{35}$
21) $6\frac{3}{35}$
22) $5\frac{6}{35}$
23) $3\frac{7}{30}$
24) $4\frac{59}{63}$
25) $2\frac{8}{45}$
26) $6\frac{16}{35}$
27) $2\frac{4}{35}$
28) $6\frac{5}{8}$
29) $2\frac{11}{21}$
30) $1\frac{3}{4}$
31) $1\frac{31}{32}$
32) $7\frac{11}{35}$
33) $2\frac{43}{72}$
34) $5\frac{7}{10}$
35) $8\frac{1}{2}$
36) $23\frac{1}{3}$
37) $3\frac{17}{26}$
38) $1\frac{39}{56}$
39) $2\frac{33}{64}$
40) $1\frac{49}{51}$
41) $1\frac{73}{75}$
42) $\frac{42}{55}$

Chapter 2:

Decimals

Math Topics that you'll learn in this Chapter:

- ✓ Comparing Decimals

- ✓ Rounding Decimals

- ✓ Adding and Subtracting Decimals

- ✓ Multiplying and Dividing Decimals

Comparing Decimals

- Decimal is a fraction written in a special form. For example, instead of writing $\frac{1}{2}$ you can write 0.5

- A Decimal Number contains a Decimal Point. It separates the whole number part from the fractional part of a decimal number.

- Let's review decimal place values: Example: 53.9861

    5: tens   3: ones   9: tenths
    8: hundredths   6: thousandths   1: tens thousandths

☑ To compare decimals, compare each digit of two decimals in the same place value. Start from left. Compare hundreds, tens, ones, tenth, hundredth, etc.

☑ To compare numbers, use these symbols:

Equal to $=$,   Less than $<$,   Greater than $>$
Greater than or equal $\geq$,   Less than or equal $\leq$

Examples:

1) Compare 0.60 and 0.06.

    Solution: 0.60 *is greater than* 0.06, because the tenth place of 0.60 is 6, but the tenth place of 0.06 is zero. Then: $0.60 > 0.06$

2) Compare 0.0815 and 0.815.

    Solution: 0.815 *is greater than* 0.0815, because the tenth place of 0.815 is 8, but the tenth place of 0.0815 is zero. Then: $0.0815 < 0.815$

Rounding Decimals

☑ We can round decimals to a certain accuracy or number of decimal places. This is used to make calculation easier to do and results easier to understand, when exact values are not too important.

☑ First, you'll need to remember your place values: For example:
$$12.4869$$

| 1: tens | 2: ones | 4: tenths |
| 8: hundredths | 6: thousandths | 9: tens thousandths |

☑ To round a decimal, first find the place value you'll round to.

☑ Find the digit to the right of the place value you're rounding to. If it is 5 or bigger, add 1 to the place value you're rounding to and remove all digits on its right side. If the digit to the right of the place value is less than 5, keep the place value and remove all digits on the right.

Examples:

1) Round 1.9278 to the thousandth place value.

   Solution: First look at the next place value to the right, (tens thousandths). It's 8 and it is greater than 5. Thus add 1 to the digit in the thousandth place. Thousandth place is 7. → $7 + 1 = 8$, then, the answer is 1.928

2) Round 9.4126 to the nearest hundredth.

   Solution: First look at the digit to the right of hundredth (thousandths place value). It's 2 and it is less than 5, thus remove all the digits to the right of hundredth place. Then, the answer is 9.41

Adding and Subtracting Decimals

✓ Line up the decimal numbers.

✓ Add zeros to have same number of digits for both numbers if necessary.

✓ Remember your place values: For example:

$$73.5196$$

- 7: tens
- 3: ones
- 5: tenths
- 1: hundredths
- 9: thousandths
- 6: tens thousandths

✓ Add or subtract using column addition or subtraction.

Examples:

1) Add. $1.8 + 3.12$

   Solution: First line up the numbers: $\begin{array}{r}1.8\\+3.12\\\hline\end{array}$ → Add a zero to have same number of digits for both numbers. $\begin{array}{r}1.80\\+3.12\\\hline\end{array}$ → Start with the hundredths place: $0 + 2 = 2$, $\begin{array}{r}1.80\\+3.12\\\hline 2\end{array}$ → Continue with tenths place: $8 + 1 = 9$, $\begin{array}{r}1.80\\+3.12\\\hline .92\end{array}$ → Add the ones place: $3 + 1 = 4$, $\begin{array}{r}1.80\\+3.12\\\hline 4.92\end{array}$

2) Find the difference. $3.67 - 2.23$

   Solution: First line up the numbers: $\begin{array}{r}3.67\\-2.23\\\hline\end{array}$ → Start with the hundredths place: $7 - 3 = 4$, $\begin{array}{r}3.67\\-2.23\\\hline 4\end{array}$ → Continue with tenths place. $6 - 2 = 4$, $\begin{array}{r}3.67\\-2.23\\\hline .44\end{array}$ → Subtract the ones place. $3 - 2 = 1$, $\begin{array}{r}3.67\\-2.23\\\hline 1.44\end{array}$

Multiplying and Dividing Decimals

For multiplying decimals:

✓ Ignore the decimal point and set up and multiply the numbers as you do with whole numbers.

✓ Count the total number of decimal places in both of the factors.

✓ Place the decimal point in the product.

For dividing decimals:

✓ If the divisor is not a whole number, move decimal point to right to make it a whole number. Do the same for dividend.

✓ Divide similar to whole numbers.

Examples:

1) Find the product. $0.81 \times 0.32 =$

   Solution: Set up and multiply the numbers as you do with whole numbers. Line up the numbers: $\frac{81}{\times 32}$ → Start with the ones place then continue with other digits → $\frac{81}{\times 32}$ / $2,592$. Count the total number of decimal places in both of the factors. There are four decimals digits. (two for each factor 0.81 and 0.32) Then: $0.81 \times 0.32 = 0.2592$

2) Find the quotient. $1.60 \div 0.4 =$

   Solution: The divisor is not a whole number. Multiply it by 10 to get 4: → $0.4 \times 10 = 4$

   Do the same for the dividend to get 16. → $1.60 \times 10 = 1.6$

   Now, divide: $16 \div 4 = 4$. The answer is 4.

Chapter 2: Practices

✎ *Compare. Use >, =, and <*

1) 0.88 ☐ 0.088
2) 0.56 ☐ 0.57
3) 0.99 ☐ 0.89
4) 1.55 ☐ 1.65
5) 1.58 ☐ 1.75
6) 2.91 ☐ 2.85

✎ *Round each decimal to the nearest whole number.*

7) 5.94
8) 16.47
9) 9.7
10) 35.8
11) 24.46
12) 12.5

✎ *Find the sum or difference.*

13) $43.15 + 23.65 =$
14) $56.74 - 22.43 =$
15) $25.47 + 31.76 =$
16) $69.87 - 35.98 =$
17) $45.53 + 18.95 =$
18) $25.13 - 18.72 =$

✎ *Find the product and quotient.*

19) $0.5 \times 0.8 =$
20) $6.4 \div 0.4 =$
21) $3.25 \times 2.2 =$
22) $8.4 \div 2.5 =$
23) $5.4 \times 0.6 =$
24) $1.42 \div 0.5 =$

Answers – Chapter 2

1) 0.88 > 0.088
2) 0.56 < 0.57
3) 0.99 > 0.89
4) 1.55 < 1.65
5) 1.58 < 1.75
6) 2.91 > 2.85
7) 6
8) 16
9) 10
10) 36
11) 24
12) 13
13) 66.8
14) 34.31
15) 57.23
16) 33.89
17) 64.48
18) 6.41
19) 0.4
20) 16
21) 7.15
22) 3.36
23) 3.24
24) 2.84

Chapter 3:

Integers and Order of Operations

Math Topics that you'll learn in this Chapter:

- ✓ Adding and Subtracting Integers
- ✓ Multiplying and Dividing Integers
- ✓ Order of Operations
- ✓ Integers and Absolute Value

Adding and Subtracting Integers

- ☑ Integers include: zero, counting numbers, and the negative of the counting numbers. $\{\ldots, -3, -2, -1, 0, 1, 2, 3, \ldots\}$
- ☑ Add a positive integer by moving to the right on the number line. (you will get a bigger number)
- ☑ Add a negative integer by moving to the left on the number line. (you will get a smaller number)
- ☑ Subtract an integer by adding its opposite.

Examples:

1) Solve. $(-4) - (-5) =$

   Solution: Keep the first number and convert the sign of the second number to its opposite. (change subtraction into addition. Then: $(-4) + 5 = 1$

2) Solve. $11 + (8 - 19) =$

   Solution: First subtract the numbers in brackets, $8 - 19 = -11$.
   Then: $11 + (-11) = \rightarrow$ change addition into subtraction: $11 - 11 = 0$

3) Solve. $5 - (-14 - 3) =$

   Solution: First subtract the numbers in brackets, $-14 - 3 = -17$
   Then: $5 - (-17) = \rightarrow$ change subtraction into addition: $5 + 17 = 22$

4) Solve. $10 + (-6 - 15) =$

   Solution: First subtract the numbers in brackets, $-6 - 15 = -21$
   Then: $10 + (-21) = \rightarrow$ change addition into subtraction: $10 - 21 = -11$

Multiplying and Dividing Integers

Use following rules for multiplying and dividing integers:

- (negative) × (negative) = positive
- (negative) ÷ (negative) = positive
- (negative) × (positive) = negative
- (negative) ÷ (positive) = negative
- (positive) × (positive) = positive
- (positive) ÷ (negative) = negative

Examples:

1) Solve. $2 \times (-3) =$

   Solution: Use this rule: (positive) × (negative) = negative.
   Then: $(2) \times (-3) = -6$

2) Solve. $(-5) + (-27 \div 9) =$

   Solution: First divided -27 by 9, the numbers in brackets, use this rule: (negative) ÷ (positive) = negative. Then: $-27 \div 9 = -3$
   $(-5) + (-27 \div 9) = (-5) + (-3) = -5 - 3 = -8$

3) Solve. $(15 - 17) \times (-8) =$

   Solution: First subtract the numbers in brackets, $15 - 17 = -2 \rightarrow (-2) \times (-8) =$

   Now use this rule: (negative) × (negative) = positive
   $(-2) \times (-8) = 16$

4) Solve. $(16 - 10) \div (-2) =$

   Solution: First subtract the numbers in brackets, $16 - 10 = 6 \rightarrow (6) \div (-2) =$

   Now use this rule: (positive) ÷ (negative) = negative
   $(6) \div (-2) = -3$

Order of Operations

- ✓ In Mathematics, "operations" are addition, subtraction, multiplication, division, exponentiation (written as b^n), and grouping;
- ✓ When there is more than one math operation in an expression, use PEMDAS: (to memorize this rule, remember the phrase "Please Excuse My Dear Aunt Sally".)
    - ❖ Parentheses
    - ❖ Exponents
    - ❖ Multiplication and Division (from left to right)
    - ❖ Addition and Subtraction (from left to right)

Examples:

1) Calculate. $(3 + 5) \div (3^2 \div 9) =$

   Solution: First simplify inside parentheses: $(8) \div (9 \div 9) = (8) \div (1)$, Then: $(8) \div (1) = 8$

2) Solve. $(7 \times 8) - (12 - 4) =$

   Solution: First calculate within parentheses: $(7 \times 8) - (12 - 4) = (56) - (8)$, Then: $(56) - (8) = 48$

3) Calculate. $-2[(8 \times 9) \div (2^2 \times 2)] =$

   Solution: First calculate within parentheses: $-2[(72) \div (4 \times 2)] = -2[(72) \div (8)] = -2[9]$ multiply -2 and 9. Then: $-2[9] = -18$

4) Solve. $(14 \div 7) + (-13 + 8) =$

   Solution: First calculate within parentheses: $(14 \div 7) + (-13 + 8) = (2) + (-5)$

   Then: $(2) - (5) = -3$

Integers and Absolute Value

- ✓ The absolute value of a number is its distance from zero, in either direction, on the number line. For example, the distance of 9 and −9 from zero on number line is 9.
- ✓ The absolute value of an integer is the numerical value without its sign. (negative or positive)
- ✓ The vertical bar is used for absolute value as in $|x|$.
- ✓ The absolute value of a number is never negative; because it only shows, "how far the number is from zero".

Examples:

1) Calculate. $|12 - 4| \times 4 =$

   Solution: First solve $|12 - 4|$, → $|12 - 4| = |8|$, the absolute value of 8 is 8, $|8| = 8$
   Then: $8 \times 4 = 32$

2) Solve. $\frac{|-16|}{4} \times |3 - 8| =$

   Solution: First find $|-16|$, → the absolute value of −16 is 16, then: $|-16| = 16$,
   $\frac{16}{4} \times |3 - 8| =$
   Now, calculate $|3 - 8|$, → $|3 - 8| = |-5|$, the absolute value of −5 is 5. $|-5| = 5$
   Then: $\frac{16}{4} \times 5 = 4 \times 5 = 20$

3) Solve. $|9 - 3| \times \frac{|-3 \times 8|}{6} =$

   Solution: First calculate $|9 - 3|$, → $|9 - 3| = |6|$, the absolute value of 6 is 6, $|6| = 6$. Then:
   $6 \times \frac{|-3 \times 8|}{6}$
   Now calculate $|-3 \times 8|$, → $|-3 \times 8| = |-24|$, the absolute value of −24 is 24, $|-24| = 24$
   Then: $6 \times \frac{24}{6} = 6 \times 4 = 24$

Chapter 3: Practices

✎ *Find each sum or difference.*

1) $18 + (-5) =$
2) $(-16) + 24 =$
3) $(-12) + (-9) =$
4) $14 + (-8) + 6 =$
5) $24 + (-10 - 7) =$
6) $(-15) + (-6 + 12) =$

✎ *Find each product or quotient.*

7) $8 \times (-6) =$
8) $(-12) \div (-3) =$
9) $(-4) \times (-7) \times 2 =$
10) $3 \times (-5) \times (-6) =$
11) $(-7 - 37) \div (-11) =$
12) $(8 - 6) \times (-24) =$

✎ *Evaluate each expression.*

13) $8 + (3 \times 7) =$
14) $(18 \times 2) - 14 =$
15) $(15 - 7) + (2 \times 6) =$
16) $(8 + 4) \div (2^3 \div 2) =$
17) $2[(6 \times 3) \div (3^2 \times 2)] =$
18) $-3[(8 \times 2^2) \div (8 \times 2)] =$

✎ *Find the answers.*

19) $|-6| + |9 - 12| =$
20) $|8| - |7 - 19| + 1 =$
21) $\frac{|-40|}{8} \times \frac{|-15|}{5} =$
22) $|7 \times -5| \times \frac{|-3|}{8} =$
23) $\frac{|-121|}{11} - |-8 \times 2| =$
24) $\frac{|-3 \times -6|}{9} \times \frac{|4 \times -6|}{8} =$

Answers – Chapter 3

1) 13
2) 8
3) −21
4) 12
5) 7
6) −9
7) −48
8) 4
9) 56
10) 90
11) 4
12) −48

13) 29
14) 22
15) 20
16) 3
17) 2
18) −6
19) 9
20) −3
21) 15
22) 140
23) −5
24) 6

Chapter 4:

Ratios and Proportions

Math Topics that you'll learn in this Chapter:

- ✓ Simplifying Ratios
- ✓ Proportional Ratios
- ✓ Similarity and Ratios

Simplifying Ratios

☑ Ratios are used to make comparisons between two numbers.
☑ Ratios can be written as a fraction, using the word "to", or with a colon. Example: $\frac{3}{4}$ or "3 to 4" or 3:4
☑ You can calculate equivalent ratios by multiplying or dividing both sides of the ratio by the same number.

Examples:

1) Simplify. $9:3 =$

   Solution: Both numbers 9 and 3 are divisible by 3, $\Rightarrow 9 \div 3 = 3$, $3 \div 3 = 1$, Then: $9:3 = 3:1$

2) Simplify. $\frac{24}{44} =$

   Solution: Both numbers 24 and 44 are divisible by 4, $\Rightarrow 24 \div 4 = 6$, $44 \div 4 = 11$, Then: $\frac{24}{44} = \frac{6}{11}$

3) There are 36 students in a class and 16 of them are girls. Write the ratio of girls to boys.

   Solution: Subtract 16 from 36 to find the number of boys in the class. $36 - 16 = 20$. There are 20 boys in the class. So, ratio of girls to boys is $16:20$. Now, simplify this ratio. Both 20 and 16 are divisible by 4. Then: $20 \div 4 = 5$, and $16 \div 4 = 4$. In simplest form, this ratio is $4:5$

4) A recipe calls for butter and sugar in the ratio $3:4$. If you're using 9 cups of butter, how many cups of sugar should you use?

   Solution: Since, you use 9 cups of butter, or 3 times as much, you need to multiply the amount of sugar by 3. Then: $4 \times 3 = 12$. So, you need to use 12 cups of sugar. You can solve this using equivalent fractions: $\frac{3}{4} = \frac{9}{12}$

Proportional Ratios

- Two ratios are proportional if they represent the same relationship.
- A proportion means that two ratios are equal. It can be written in two ways: $\frac{a}{b} = \frac{c}{d}$ $a : b = c : d$
- The proportion $\frac{a}{b} = \frac{c}{d}$ can be written as: $a \times d = c \times b$

Examples:

1) Solve this proportion for x. $\frac{3}{7} = \frac{12}{x}$

   Solution: Use cross multiplication: $\frac{3}{7} = \frac{12}{x} \Rightarrow 3 \times x = 7 \times 12 \Rightarrow 3x = 84$

   Divide both sides by 3 to find x: $x = \frac{84}{3} \Rightarrow x = 28$

2) If a box contains red and blue balls in ratio of $3:7$ red to blue, how many red balls are there if 49 blue balls are in the box?

   Solution: Write a proportion and solve. $\frac{3}{7} = \frac{x}{49}$

   Use cross multiplication: $3 \times 49 = 7 \times x \Rightarrow 147 = 7x$

   Divide to find x: $x = \frac{147}{7} \Rightarrow x = 21$. There are 21 red balls in the box.

3) Solve this proportion for x. $\frac{2}{9} = \frac{12}{x}$

   Solution: Use cross multiplication: $\frac{2}{9} = \frac{12}{x} \Rightarrow 2 \times x = 9 \times 12 \Rightarrow 2x = 108$

   Divide to find x: $x = \frac{108}{2} \Rightarrow x = 54$

4) Solve this proportion for x. $\frac{6}{7} = \frac{18}{x}$

   Solution: Use cross multiplication: $\frac{6}{7} = \frac{18}{x} \Rightarrow 6 \times x = 7 \times 18 \Rightarrow 6x = 126$

   Divide to find x: $x = \frac{126}{6} \Rightarrow x = 21$

Similarity and Ratios

✓ Two figures are similar if they have the same shape.

✓ Two or more figures are similar if the corresponding angles are equal, and the corresponding sides are in proportion.

Examples:

1) Following triangles are similar. What is the value of unknown side?

   Solution: Find the corresponding sides and write a proportion.
   $\frac{5}{10} = \frac{4}{x}$. Now, use cross product to solve for x:
   $\frac{5}{10} = \frac{4}{x} \to 5 \times x = 10 \times 4 \to 5x = 40$. Divide both sides by 5. Then: $5x = 40 \to \frac{5x}{5} = \frac{40}{5} \to x = 8$

   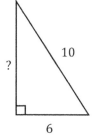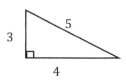

   The missing side is 8.

2) Two rectangles are similar. The first is 6 feet wide and 20 feet long. The second is 15 feet wide. What is the length of the second rectangle?

   Solution: Let's put x for the length of the second rectangle. Since two rectangles are similar, their corresponding sides are in proportion. Write a proportion and solve for the missing number. $\frac{6}{15} = \frac{20}{x} \to 6x = 15 \times 20 \to 6x = 300 \to x = \frac{300}{6} = 50$

   The length of the second rectangle is 50 feet.

Chapter 4: Practices

✎ *Reduce each ratio.*

1) $9:18 =$ ___ : ___
2) $6:54 =$ ___ : ___
3) $28:49 =$ ___ : ___
4) Bob has 12 red cards and 20 green cards. What is the ratio of Bob's red cards to his green cards? _____

5) In a party, 10 soft drinks are required for every 12 guests. If there are 252 guests, how many soft drinks is required? _____

6) In Jack's class, 18 of the students are tall and 10 are short. In Michael's class 54 students are tall and 30 students are short. Which class has a higher ratio of tall to short students? _____

✎ *Solve each proportion.*

7) $\frac{3}{7} = \frac{18}{x}$, $x =$ ____

8) $\frac{5}{9} = \frac{x}{108}$, $x =$ ____

9) $\frac{2}{13} = \frac{8}{x}$, $x =$ ____

10) $\frac{4}{10} = \frac{6}{x}$, $x =$ ____

11) $\frac{8}{20} = \frac{x}{65}$, $x =$ ____

12) $\frac{6}{15} = \frac{14}{x}$, $x =$ ____

✎ *Solve each problem.*

13) Two rectangles are similar. The first is 8 *feet* wide and 22 *feet* long. The second is 12 *feet* wide. What is the length of the second rectangle? _____

14) Two rectangles are similar. One is 3.2 *meters* by 8 *meters*. The longer side of the second rectangle is 34.5 *meters*. What is the other side of the second rectangle? _____

Answers – Chapter 4

1) 1 : 2
2) 1 : 9
3) 4 : 7
4) 3 : 5
5) 210
6) The ratio for both classes is 9 to 5.
7) 42
8) 60
9) 52
10) 15
11) 26
12) 35
13) 33 feet
14) 13.8 meters

Chapter 5:

Percentage

Math Topics that you'll learn in this Chapter:

- ✓ Percentage Calculations

- ✓ Percent Problems

- ✓ Percent of Increase and Decrease

- ✓ Discount, Tax and Tip

- ✓ Simple Interest

Percent Problems

- ✓ Percent is a ratio of a number and 100. It always has the same denominator, 100. Percent symbol is "%".
- ✓ Percent means "per 100". So, 20% is 20/100.
- ✓ In each percent problem, we are looking for the base, or part or the percent.
- ✓ Use the following equations to find each missing section in a percent problem:
    - Base = Part ÷ Percent
    - Part = Percent × Base
    - Percent = Part ÷ Base

Examples:

1) What is 25% of 60?

   Solution: In this problem, we have percent (25%) and base (60) and we are looking for the "part". Use this formula: $part = percent \times base$. Then: $part = 25\% \times 60 = \frac{25}{100} \times 60 = 0.25 \times 60 = 15$.

   The answer:

   25% of 60 is 15.

2) 20 is what percent of 400?

   Solution: In this problem, we are looking for the percent. Use this equation:

   $Percent = Part \div Base \rightarrow Percent = 20 \div 400 = 0.05 = 5\%$.

   Then: 20 is 5 percent of 400.

Percent of Increase and Decrease

☑ Percent of change (increase or decrease) is a mathematical concept that represents the degree of change over time.

☑ To find the percentage of increase or decrease:

1. New Number − Original Number
2. The result ÷ Original Number × 100

☑ Or use this formula: Percent of change = $\frac{new\ number - original\ number}{original\ number} \times 100$

☑ Note: If your answer is a negative number, then this is a percentage decrease. If it is positive, then this is a percentage increase.

Examples:

1) The price of a shirt increases from $20 to $30. What is the percentage increase?
   Solution: First find the difference: 30 − 20 = 10
   Then: $10 \div 20 \times 100 = \frac{10}{20} \times 100 = 50$. The percentage increase is 50. It means that the price of the shirt increased 50%.

2) The price of a table increased from $25 to $40. What is the percent of increase?
   Solution: Use percentage formula: Percent of change = $\frac{new\ number - original\ number}{original\ number} \times 100 = \frac{40-25}{25} \times 100 = \frac{15}{25} \times 100 = 0.6 \times 100 = 60$. The percentage increase is 60. It means that the price of the table increased 60%.

3) The population of a town was 50,000 in the 2000 census and 40,000 in the 2010 census. By what percent did the population decrease?
   Solution: Use percentage formula:
   Percent of change = $\frac{new\ number - original\ number}{original\ number} \times 100 = \frac{40,000-50,000}{50,000} \times 100 = \frac{-10,000}{50,000} \times 100 = -0.2 \times 100 = -20$. The population of the town decreased by 20%.

Discount, Tax and Tip

- ✓ To find discount: Multiply the regular price by the rate of discount
- ✓ To find selling price: Original price – discount
- ✓ To find tax: Multiply the tax rate to the taxable amount (income, property value, etc.)
- ✓ To find tip, multiply the rate to the selling price.

Examples:

1) With an 10% discount, Ella was able to save $45 on a dress. What was the original price of the dress?

   Solution: let x be the original price of the dress. Then: $10\% \; of \; x = 45$. Write an equation and solve for x: $0.10 \times x = 45 \rightarrow x = \frac{45}{0.10} = 450$. The original price of the dress was $450.

2) Sophia purchased a new computer for a price of $950 at the Apple Store. What is the total amount her credit card is charged if the sales tax is 7%?

   Solution: The taxable amount is $950, and the tax rate is 7%. Then: $Tax = 0.07 \times 950 = 66.50$

   $Final \; price = Selling \; price + Tax \rightarrow final \; price = \$950 + \$66.50 = \$1,016.50$

3) Nicole and her friends went out to eat at a restaurant. If their bill was $80.00 and they gave their server a 15% tip, how much did they pay altogether?

   Solution: First find the tip. To find tip, multiply the rate to the bill amount.
   $Tip = 80 \times 0.15 = 12$. The final price is: $\$80 + \$12 = \$92$

Simple Interest

- ✓ Simple Interest: The charge for borrowing money or the return for lending it.
- ✓ Simple interest is calculated on the initial amount (principal).
- ✓ To solve a simple interest problem, use this formula:

  Interest = principal × rate × time   ($I = p \times r \times t = prt$)

Examples:

1) Find simple interest for $300 investment at 6% for 5 years.

   Solution: Use Interest formula: $I = prt$ ($P = \$300$, r = 6% = $\frac{6}{100}$ = 0.06 and $t = 5$)
   Then: $I = 300 \times 0.06 \times 5 = \$90$

2) Find simple interest for $1,600 at 5% for 2 years.

   Solution: Use Interest formula: $I = prt$ ($P = \$1,600$, r = 5% = $\frac{5}{100}$ = 0.05 and $t = 2$)
   Then: $I = 1,600 \times 0.05 \times 2 = \$160$

3) Andy received a student loan to pay for his educational expenses this year. What is the interest on the loan if he borrowed $6,500 at 8% for 6 years?
   Solution: Use Interest formula: $I = prt$. $P = \$6,500$, r = 8% = 0.08 and $t = 6$
   Then: $I = 6,500 \times 0.08 \times 8 = \$3,120$

4) Bob is starting his own small business. He borrowed $10,000 from the bank at a 6% rate for 6 months. Find the interest Bob will pay on this loan.
   Solution: Use Interest formula: $I = prt$. $P = \$10,000$, r = 6% = 0.06 and $t = 0.5$ (6 months is half year). Then: $I = 10,000 \times 0.06 \times 0.5 = \$300$

Chapter 5: Practices

✍ *Solve each problem.*

1) 15 is what percent of 60? ____%

2) 18 is what percent of 24? ____%

3) 25 is what percent of 500? ____%

4) 14 is what percent of 280? ____%

5) 45 is what percent of 180? ____%

6) 70 is what percent of 350? ____%

✍ *Solve each percent of change word problem.*

7) Bob got a raise, and his hourly wage increased from $15 to $18. What is the percent increase? _____ %

8) The price of a pair of shoes increases from $40 to $66. What is the percent increase? ____ %

9) At a coffeeshop, the price of a cup of coffee increased from $1.20 to $1.44. What is the percent increase in the cost of the coffee? _____ %

✍ *Find the selling price of each item.*

10) Original price of a computer: $650

Tax: 8%, Selling price: $_____

11) Original price of a laptop: $480

Tax: 15%, Selling price: $_____

12) Nicolas hired a moving company. The company charged $400 for its services, and Nicolas gives the movers a 15% tip. How much does Nicolas tip the movers? $_____

13) Mason has lunch at a restaurant and the cost of his meal is $30. Mason wants to leave a 20% tip. What is Mason's total bill including tip? $_____

✎ *Determine the simple interest for these loans.*

14) $440 at 5% for 6 years. $___

15) $460 at 2.5% for 4 years. $___

16) $500 at 3% for 5 years. $___

17) $550 at 9% for 2 years. $___

18) A new car, valued at $28,000, depreciates at 9% per year. What is the value of the car one year after purchase? $_____

19) Sara puts $4,000 into an investment yielding 5% annual simple interest; she left the money in for five years. How much interest does Sara get at the end of those five years? $_____

Answers – Chapter 5

1) 25%
2) 75%
3) 5%
4) 5%
5) 25%
6) 20%
7) 20%
8) 65%
9) 20%
10) $702.00
11) $552.00
12) $60.00
13) $36.00
14) $132
15) $46
16) $75
17) $99
18) $25,480.00
19) $1,000.00

Chapter 6:

Expressions and Variables

Math Topics that you'll learn in this Chapter:

- ✓ Simplifying Variable Expressions
- ✓ Simplifying Polynomial Expressions
- ✓ The Distributive Property
- ✓ Evaluating One Variable
- ✓ Evaluating Two Variables

Simplifying Variable Expressions

- In algebra, a variable is a letter used to stand for a number. The most common letters are: $x, y, z, a, b, c, m,$ and n.
- Algebraic expression is an expression contains integers, variables, and the math operations such as addition, subtraction, multiplication, division, etc.
- In an expression, we can combine "like" terms. (values with same variable and same power)

Examples:

1) Simplify. $(2x + 3x + 4) =$

   Solution: In this expression, there are three terms: $2x, 3x$, and 4. Two terms are "like terms": $2x$ and $3x$. Combine like terms. $2x + 3x = 5x$. Then: $(2x + 3x + 4) = 5x + 4$ (remember you cannot combine variables and numbers.)

2) Simplify. $12 - 3x^2 + 5x + 4x^2 =$

   Solution: Combine "like" terms: $-3x^2 + 4x^2 = x^2$. Then:
   $12 - 3x^2 + 5x + 4x^2 = 12 + x^2 + 5x$. Write in standard form (biggest powers first):
   $12 + x^2 + 5x = x^2 + 5x + 12$

3) Simplify. $(10x^2 + 2x^2 + 3x) =$

   Solution: Combine like terms. Then: $(10x^2 + 2x^2 + 3x) = 12x^2 + 3x$

4) Simplify. $15x - 3x^2 + 9x + 5x^2 =$

   Solution: Combine "like" terms: $15x + 9x = 24x$, and $-3x^2 + 5x^2 = 2x^2$

   Then: $15x - 3x^2 + 9x + 5x^2 = 24x + 2x^2$. Write in standard form (biggest powers first): $24x + 2x^2 = 2x^2 + 24x$

Simplifying Polynomial Expressions

☑ In mathematics, a polynomial is an expression consisting of variables and coefficients that involves only the operations of addition, subtraction, multiplication, and non-negative integer exponents of variables. $P(x) = a_n x^n + a_{n-1} x^{n-1} + \ldots + a_2 x^2 + a_1 x + a_0$

☑ Polynomials must always be simplified as much as possible. It means you must add together any like terms. (values with same variable and same power)

Examples:

1) Simplify this Polynomial Expressions. $x^2 - 5x^3 + 2x^4 - 4x^3$

   Solution: Combine "like" terms: $-5x^3 - 4x^3 = -9x^3$

   Then: $x^2 - 5x^3 + 2x^4 - 4x^3 = x^2 - 9x^3 + 2x^4$

   Now, write the expression in standard form: $2x^4 - 9x^3 + x^2$

2) Simplify this expression. $(2x^2 - x^3) - (x^3 - 4x^2) =$

   Solution: First use distributive property: → multiply $(-)$ into $(x^3 - 4x^2)$

   $(2x^2 - x^3) - (x^3 - 4x^2) = 2x^2 - x^3 - x^3 + 4x^2$

   Then combine "like" terms: $2x^2 - x^3 - x^3 + 4x^2 = 6x^2 - 2x^3$

   And write in standard form: $6x^2 - 2x^3 = -2x^3 + 6x^2$

3) Simplify. $4x^4 - 5x^3 + 15x^4 - 12x^3 =$

   Solution: Combine "like" terms: $-5x^3 - 12x^3 = -17x^3$ and $4x^4 + 15x^4 = 19x^4$

   Then: $4x^4 - 5x^3 + 15x^4 - 12x^3 = 19x^4 - 17x^3$

The Distributive Property

- ☑ The distributive property (or the distributive property of multiplication over addition and subtraction) simplifies and solves expressions in the form of: $a(b + c)$ or $a(b - c)$
- ☑ The distributive property is multiplying a term outside the parentheses by the terms inside.
- ☑ Distributive Property rule: $a(b + c) = ab + ac$

Examples:

1) *Simply using distributive property.* $(-4)(x - 5)$

   Solution: Use Distributive Property rule: $a(b + c) = ab + ac$
   $(-4)(x - 5) = (-4 \times x) + (-4) \times (-5) = -4x + 20$

2) *Simply.* $(3)(2x - 4)$

   Solution: Use Distributive Property rule: $a(b + c) = ab + ac$
   $(3)(2x - 4) = (3 \times 2x) + (3) \times (-4) = 6x - 12$

3) *Simply.* $(-3)(3x - 5) + 4x$

   Solution: First, simplify $(-3)(3x - 5)$ using distributive property.
   Then: $(-3)(3x - 5) = -9x + 15$
   Now combine like terms: $(-3)(3x - 5) + 4x = -9x + 15 + 4x$
   In this expression, $-9x$ and $4x$ are "like terms" and we can combine them.
   $-9x + 4x = -5x$. Then: $-9x + 15 + 4x = -5x + 15$

Evaluating One Variable

☑ To evaluate one variable expressions, find the variable and substitute a number for that variable.

☑ Perform the arithmetic operations.

Examples:

1) *Calculate this expression for* $x = 3$. $15 - 3x$

    Solution: First substitute 3 for x

    Then: $15 - 3x = 15 - 3(3)$

    Now, use order of operation to find the answer: $15 - 3(3) = 15 - 9 = 6$

2) *Evaluate this expression for* $x = 1$. $5x - 12$

    Solution: First substitute 1 for x, then:

    $5x - 12 = 5(1) - 12$

    Now, use order of operation to find the answer: $5(1) - 12 = 5 - 12 = -7$

3) *Find the value of this expression when* $x = 5$. $25 - 4x$

    Solution: First substitute 5 for x, then:

    $25 - 4x = 25 - 4(5) = 25 - 20 = 5$

4) *Solve this expression for* $x = -2$. $12 + 3x$

    Solution: Substitute -2 for x, then: $12 + 3x = 12 + 3(-2) = 12 - 6 = 6$

Evaluating Two Variables

✓ To evaluate an algebraic expression, substitute a number for each variable.

✓ Perform the arithmetic operations to find the value of the expression.

Examples:

1) *Calculate this expression for* $a = 3$ *and* $b = -2$. $3a - 6b$

   Solution: First substitute 3 for a, and -2 for b, then:
   $$3a - 6b = 3(3) - 6(-2)$$
   Now, use order of operation to find the answer: $3(3) - 6(-2) = 9 + 12 = 21$

2) *Evaluate this expression for* $x = 3$ *and* $y = 1$. $3x + 5y$

   Solution: Substitute 3 for x, and 1 for y, then:
   $$3x + 5y = 3(3) + 5(1) = 9 + 5 = 14$$

3) *Find the value of this expression when* $a = 1$ *and* $b = 2$. $5(3a - 2b)$

   Solution: Substitute 1 for a, and 2 for b, then:
   $$5(3a - 2b) = 15a - 10b = 15(1) - 10(2) = 15 - 20 = -5$$

4) *Solve this expression.* $4x - 3y$, $x = 3$, $y = 5$

   Solution: Substitute 3 for x, and 5 for y and simplify. Then: $4x - 3y = 4(3) - 3(5) = 12 - 15 = -3$

Chapter 6: Practices

✎ *Simplify each expression.*

1) $(6x - 4x + 8 + 6) =$

2) $(-14x + 26x - 12) =$

3) $(24x - 6 - 18x + 3) =$

4) $5 + 8x^2 - 9 =$

5) $7x - 4x^2 + 6x =$

6) $15x^2 - 3x - 6x^2 + 4 =$

✎ *Simplify each polynomial.*

7) $2x^2 + 5x^3 - 7x^2 + 12x =$ _____

8) $2x^4 - 5x^5 + 8x^4 - 8x^2 =$ _____

9) $5x^3 + 15x - x^2 - 2x^3 =$ _____

10) $(8x^3 - 6x^2) + (9x^2 - 10x) =$ _____

11) $(12x^4 + 4x^3) - (8x^3 - 2x^4) =$ _____

12) $(9x^5 - 7x^3) - (5x^3 + x^2) =$ _____

✎ *Use the distributive property to simply each expression.*

13) $4(5 + 6x) =$

14) $5(8 - 4x) =$

15) $(-6)(2 - 9x) =$

16) $(-7)(6x - 4) =$

17) $(3x + 12)4 =$

18) $(8x - 5)(-3) =$

✎ *Evaluate each expression using the value given.*

19) $8 - x, x = -3$

20) $x + 12, x = -6$

21) $5x - 3, x = 2$

22) $4 - 6x, x = 1$

23) $3x + 1, x = -2$

24) $15 - 2x, x = 5$

www.EffortlessMath.com

53

✏️ *Evaluate each expression using the values given.*

25) $4x - 2y$, $x = 4, y = -2$

26) $6a + 3b$, $a = 2, b = 4$

27) $12x - 5y - 8$, $x = 2, y = 3$

28) $-7a + 3b + 9$, $a = 4, b = 6$

29) $2x + 14 + 4y$, $x = 6, y = 8$

30) $4a - (5a - b) + 5$, $a = 4, b = 6$

Answers – Chapter 6

1) $2x + 14$
2) $12x - 12$
3) $6x - 3$
4) $8x^2 - 4$
5) $-4x^2 + 13x$
6) $9x^2 - 3x + 4$

7) $5x^3 - 5x^2 + 12x$
8) $-5x^5 + 10x^4 - 8x^2$
9) $3x^3 - x^2 + 15x$
10) $8x^3 + 3x^2 - 10x$
11) $14x^4 - 4x^3$
12) $9x^5 - 12x^3 - x^2$

13) $24x + 20$
14) $-20x + 40$
15) $54x - 12$
16) $-42x + 28$
17) $12x + 48$
18) $-24x + 15$

19) 11
20) 6
21) 7
22) -2
23) -5
24) 5

25) 20
26) 24
27) 1
28) -1
29) 58
30) 7

Chapter 7:

Equations and Inequalities

Math Topics that you'll learn in this Chapter:

- ✓ One–Step Equations
- ✓ Multi–Step Equations
- ✓ System of Equations
- ✓ Graphing Single–Variable Inequalities
- ✓ One–Step Inequalities
- ✓ Multi–Step Inequalities

One–Step Equations

- ☑ The values of two expressions on both sides of an equation are equal. Example: $ax = b$. In this equation, ax is equal to b.

- ☑ Solving an equation means finding the value of the variable.

- ☑ You only need to perform one Math operation in order to solve the one-step equations.

- ☑ To solve one-step equation, find the inverse (opposite) operation is being performed.

- ☑ The inverse operations are:
  - Addition and subtraction
  - Multiplication and division

Examples:

1) *Solve this equation for x.* $3x = 18, x = ?$

   Solution: Here, the operation is multiplication (variable x is multiplied by 3) and its inverse operation is division. To solve this equation, divide both sides of equation by 3:
   $$3x = 18 \rightarrow \frac{3x}{3} = \frac{18}{3} \rightarrow x = 6$$

2) *Solve this equation.* $x + 15 = 0, x = ?$

   Solution: In this equation 15 is added to the variable x. The inverse operation of addition is subtraction. To solve this equation, subtract 15 from both sides of the equation: $x + 15 - 15 = 0 - 15$. Then simplify: $x + 15 - 15 = 0 - 15 \rightarrow x = -15$

3) *Solve this equation for x.* $x - 23 = 0$

   Solution: Here, the operation is subtraction and its inverse operation is addition. To solve this equation, add 23 to both sides of the equation: $x + 23 - 23 = 0 - 23 \rightarrow x = -23$

Multi-Step Equations

- ✓ To solve a multi-step equation, combine "like" terms on one side.
- ✓ Bring variables to one side by adding or subtracting.
- ✓ Simplify using the inverse of addition or subtraction.
- ✓ Simplify further by using the inverse of multiplication or division.
- ✓ Check your solution by plugging the value of the variable into the original equation.

Examples:

1) Solve this equation for x. $3x + 6 = 16 - 2x$

   Solution: First bring variables to one side by adding $2x$ to both sides. Then:

   $3x + 6 = 16 - 2x \to 3x + 6 + 2x = 16 - 2x + 2x$. Simplify: $5x + 6 = 16$

   Now, subtract 6 from both sides of the equation: $5x + 6 - 6 = 16 - 6 \to 5x = 10 \to$

   Divide both sides by 5: $5x = 10 \to \frac{5x}{5} = \frac{10}{5} \to x = 2$

   Let's check this solution by substituting the value of 2 for x in the original equation:

   $x = 2 \to 3x + 6 = 16 - 2x \to 3(2) + 6 = 16 - 2(2) \to 6 + 6 = 16 - 4 \to 12 = 12$

   The answer $x = 2$ is correct.

2) Solve this equation for x. $-4x + 4 = 16$

   Solution: Subtract 4 from both sides of the equation. $-4x + 4 - 4 = 16 - 4 \to -4x = 12$

   Divide both sides by -4, then: $-4x = 12 \to \frac{-4x}{-4} = \frac{12}{-4} \to x = -3$

   Now, check the solution: $x = -3 \to -4x + 4 = 16 \to -4(-3) + 4 = 16 \to 16 = 16$

   The answer $x = -2$ is correct.

System of Equations

✓ A system of equations contains two equations and two variables. For example, consider the system of equations: $x - y = 1, x + y = 5$

✓ The easiest way to solve a system of equations is using the elimination method. The elimination method uses the addition property of equality. You can add the same value to each side of an equation.

✓ For the first equation above, you can add $x + y$ to the left side and 5 to the right side of the first equation: $x - y + (x + y) = 1 + 5$. Now, if you simplify, you get: $x - y + (x + y) = 1 + 5 \rightarrow 2x = 6 \rightarrow x = 3$. Now, substitute 3 for the x in the first equation: $3 - y = 1$. By solving this equation, $y = 2$

Example:

What is the value of $x + y$ in this system of equations? $\begin{cases} x + 2y = 6 \\ 2x - y = -8 \end{cases}$

Solution: Solving a System of Equations by Elimination:

Multiply the first equation by (-2), then add it to the second equation.

$\begin{array}{l} -2(x + 2y = 6) \\ 2x - y = -8 \end{array} \Rightarrow \begin{array}{l} -2x - 4y = -12 \\ 2x - y = -8 \end{array} \Rightarrow -5y = -20 \Rightarrow y = 4$

Plug in the value of y into one of the equations and solve for x.

$x + 2(4) = 6 \Rightarrow x + 8 = 6 \Rightarrow x = 6 - 8 \Rightarrow x = -2$

Thus, $x + y = -2 + 4 = 2$

Graphing Single–Variable Inequalities

- An inequality compares two expressions using an inequality sign.
- Inequality signs are: "less than" <, "greater than" >, "less than or equal to" ≤, and "greater than or equal to" ≥.
- To graph a single-variable inequality, find the value of the inequality on the number line.
- For less than (<) or greater than (>) draw open circle on the value of the variable. If there is an equal sign too, then use filled circle.
- Draw an arrow to the right for greater or to the left for less than.

Examples:

1) Draw a graph for this inequality. $x > 3$

Solution: Since, the variable is greater than 3, then we need to find 3 in the number line and draw an open circle on it.

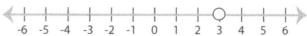

Then, draw an arrow to the right.

2) Graph this inequality. $x \leq -4$.

Solution: Since, the variable is less than or equal to −4, then we need to find −4 in the number line and draw a filled circle on it. Then, draw an arrow to the left.

One–Step Inequalities

- An inequality compares two expressions using an inequality sign.
- Inequality signs are: "less than" <, "greater than" >, "less than or equal to" ≤, and "greater than or equal to" ≥.
- You only need to perform one Math operation in order to solve the one-step inequalities.
- To solve one-step inequalities, find the inverse (opposite) operation is being performed.
- For dividing or multiplying both sides by negative numbers, flip the direction of the inequality sign.

Examples:

1) Solve this inequality for x. $x + 3 \geq 4$

   Solution: The inverse (opposite) operation of addition is subtraction. In this inequality, 3 is added to x. To isolate x we need to subtract 3 from both sides of the inequality. Then:

   $x + 3 \geq 4 \rightarrow x + 3 - 3 \geq 4 - 3 \rightarrow x \geq 1$. The solution is: $x \geq 1$

2) Solve the inequality. $x - 5 > -4$.

   Solution: 5 is subtracted from x. Add 5 to both sides. $x - 5 > -4 \rightarrow x - 5 + 5 > -4 + 5 \rightarrow x > 1$

3) Solve. $2x \leq -4$.

   Solution: 2 is multiplied to x. Divide both sides by 2. Then: $2x \leq -4 \rightarrow \frac{2x}{2} \leq \frac{-4}{2} \rightarrow x \leq -2$

4) Solve. $-6x \leq 12$.

   Solution: -6 is multiplied to x. Divide both sides by -6. Remember when dividing or multiplying both sides of an inequality by negative numbers, flip the direction of the inequality sign. Then:

   $$-6x \leq 12 \rightarrow \frac{-6x}{-6} \geq \frac{12}{-6} \rightarrow x \geq -2$$

Multi-Step Inequalities

- To solve a multi-step inequality, combine "like" terms on one side.
- Bring variables to one side by adding or subtracting.
- Isolate the variable.
- Simplify using the inverse of addition or subtraction.
- Simplify further by using the inverse of multiplication or division.
- For dividing or multiplying both sides by negative numbers, flip the direction of the inequality sign.

Examples:

1) *Solve this inequality.* $2x - 3 \leq 5$

   Solution: In this inequality, 3 is subtracted from $2x$. The inverse of subtraction is addition. Add 3 to both sides of the inequality: $2x - 3 + 3 \leq 5 + 3 \rightarrow 2x \leq 8$

   Now, divide both sides by 2. Then: $2x \leq 8 \rightarrow \frac{2x}{2} \leq \frac{8}{2} \rightarrow x \leq 4$

   The solution of this inequality is $x \leq 4$.

2) *Solve this inequality.* $3x + 9 < 12$

   Solution: First subtract 9 from both sides: $3x + 9 - 9 < 12 - 9$

   Then simplify: $3x + 9 - 9 < 12 - 9 \rightarrow 3x < 3$

   Now divide both sides by 3: $\frac{3x}{3} < \frac{3}{3} \rightarrow x < 1$

3) *Solve this inequality.* $-2x + 4 \geq 6$

   First subtract 4 from both sides: $-2x + 4 - 4 \geq 6 - 4 \rightarrow -2x \geq 2$

   Divide both sides by -2. Remember that you need to flip the direction of inequality sign.

   $$-2x \geq 2 \rightarrow \frac{-2x}{-2} \leq \frac{2}{-2} \rightarrow x \leq -1$$

Chapter 7: Practices

✎ **Solve each equation. (One–Step Equations)**

1) $x + 7 = 6, x = $ ____

2) $8 = 2 - x, x = $ ____

3) $-10 = 8 + x, x = $ ____

4) $x - 5 = -1, x = $ ____

5) $16 = x + 9, x = $ ____

6) $12 - x = -5, x = $ ____

✎ **Solve each equation. (Multi–Step Equations)**

7) $5(x + 3) = 20$

8) $-4(7 - x) = 16$

9) $8 = -2(x + 5)$

10) $14 = 3(4 - 2x)$

11) $5(x + 7) = -10$

12) $-2(6 + 3x) = 12$

✎ **Solve each system of equations.**

13) $-5x + y = -3$ $x =$
    $3x - 8y = 24$ $y =$

14) $3x - 2y = 2$ $x =$
    $x - y = 2$ $y =$

15) $4x + 7y = 2$ $x =$
    $6x + 7y = 10$ $y =$

16) $5x + 7y = 18$ $x =$
    $-3x + 7y = -22$ $y =$

✎ **Draw a graph for each inequality.**

17) $x \leq -2$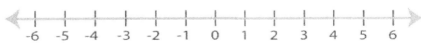

18) $x > -6$

www.EffortlessMath.com

✎ **Solve each inequality and graph it.**

19) $x - 3 \geq -1$

20) $3x - 2 < 16$

✎ **Solve each inequality.**

21) $3x + 15 > -6$

22) $-18 + 4x \leq 10$

23) $4(x + 5) \geq 8$

24) $7x - 16 < 12$

25) $3(9 + x) \geq 15$

26) $-8 + 6x > 22$

Answers – Chapter 7

1) -1
2) -6
3) -18

4) 4
5) 7
6) 17

7) 1
8) 11
9) -9

10) $-\frac{1}{3}$
11) -9
12) -4

13) $x = 0, y = -3$
14) $x = -2, y = -4$
15) $x = 4, y = -2$
16) $x = 5, y = -1$

17)

18)

19)

20)

21) $x > -7$
22) $x \leq 7$
23) $x \geq -3$
24) $x < 4$
25) $x \geq -4$
26) $x > 5$

Chapter 8:

Lines and Slope

Math Topics that you'll learn in this Chapter:

- ✓ Finding Slope

- ✓ Graphing Lines Using Slope–Intercept Form

- ✓ Writing Linear Equations

- ✓ Graphing Linear Inequalities

- ✓ Finding Midpoint

- ✓ Finding Distance of Two Points

Finding Slope

- ✓ The slope of a line represents the direction of a line on the coordinate plane.
- ✓ A coordinate plane contains two perpendicular number lines. The horizontal line is x and the vertical line is y. The point at which the two axes intersect is called the origin. An ordered pair (x, y) shows the location of a point.
- ✓ A line on coordinate plane can be drawn by connecting two points.
- ✓ To find the slope of a line, we need the equation of the line or two points on the line.
- ✓ The slope of a line with two points A (x_1, y_1) and B (x_2, y_2) can be found by using this formula: $\frac{y_2 - y_1}{x_2 - x_1} = \frac{rise}{run}$
- ✓ The equation of a line is typically written as $y = mx + b$ where m is the slope and b is the y-intercept.

Examples:

1) *Find the slope of the line through these two points*: $A(2, -7)$ and $B(4, 3)$.

    Solution: Slope $= \frac{y_2 - y_1}{x_2 - x_1}$. Let (x_1, y_1) be $A(2, -7)$ and (x_2, y_2) be $B(4, 3)$. (Remember, you can choose any point for (x_1, y_1) and (x_2, y_2)). Then: slope $= \frac{y_2 - y_1}{x_2 - x_1} = \frac{3 - (-7)}{4 - 2} = \frac{10}{2} = 5$

    The slope of the line through these two points is 5.

2) *Find the slope of the line with equation* $y = 3x + 6$

    Solution: when the equation of a line is written in the form of $y = mx + b$, the slope is m. In this line: $y = 3x + 6$, the slope is 3.

Graphing Lines Using Slope–Intercept Form

- Slope-intercept form of a line: given the slope m and the y-intercept (the intersection of the line and y-axis) b, then the equation of the line is: $y = mx + b$
- To draw the graph of a linear equation in slope-intercept form on the xy coordinate plane, find two points on the line by plugging two values for x and calculating the values of y.
- You can also use the slope (m) and one point to graph the line.

Example:

1) Sketch the graph of $y = 2x - 4$.

Solution: To graph this line, we need to find two points. When x is zero the value of y is -4. And when x is 2 the value of y is 0.

$$x = 0 \rightarrow y = 2(0) - 4 = -4,$$
$$y = 0 \rightarrow 0 = 2x - 4 \rightarrow x = 2$$

Now, we have two points: $(0, -4)$ and $(2, 0)$.

Find the points on the coordinate plane and graph the line. Remember that the slope of the line is 2.

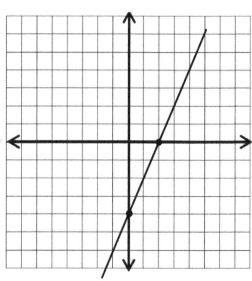

Writing Linear Equations

- ✓ The equation of a line in slope-intercept form: $y = mx + b$
- ✓ To write the equation of a line, first identify the slope.
- ✓ Find the y-intercept. This can be done by substituting the slope and the coordinates of a point (x, y) on the line.

Examples:

1) What is the equation of the line that passes through $(2, -4)$ and has a slope of 8?
   Solution: The general slope-intercept form of the equation of a line is $y = mx + b$, where m is the slope and b is the y-intercept.
   By substitution of the given point and given slope: $y = mx + b \rightarrow -4 = (2)(8) + b$
   So, $b = -4 - 16 = -20$, and the required equation is $y = 8x - 20$

2) Write the equation of the line through two points $A(2, 1)$ and $B(-2, 5)$.
   Solution: Fist find the slope: $Slop = \frac{y_2 - y_1}{x_2 - x_1} = \frac{5 - 1}{-2 - 2} = \frac{4}{-4} = -1 \rightarrow m = -1$
   To find the value of b, use either points and plug in the values of x and y in the equation. The answer will be the same: $y = -x + b$. Let's check both points. Then:
   $(2, 1) \rightarrow y = mx + b \rightarrow 1 = -1(2) + b \rightarrow b = 3$
   $(-2, 5) \rightarrow y = mx + b \rightarrow 5 = -1(-2) + b \rightarrow b = 3$. The y-intercept of the line is 3.
   The equation of the line is: $y = -x + 3$

3) What is the equation of the line that passes through $(2, -1)$ and has a slope of 5?
   Solution: The general slope-intercept form of the equation of a line is $y = mx + b$, where m is the slope and b is the y-intercept.
   By substitution of the given point and given slope: $y = mx + b \rightarrow -1 = (5)(2) + b$
   So, $b = -1 - 10 = -11$, and the equation of the line is: $y = 5x - 11$.

Finding Midpoint

- The middle of a line segment is its midpoint.
- The Midpoint of two endpoints A (x_1, y_1) and B (x_2, y_2) can be found using this formula: $M\left(\frac{x_1+x_2}{2}, \frac{y_1+y_2}{2}\right)$

Examples:

1) Find the midpoint of the line segment with the given endpoints. $(1, -3), (3, 7)$

   Solution: Midpoint = $\left(\frac{x_1+x_2}{2}, \frac{y_1+y_2}{2}\right) \rightarrow (x_1, y_1) = (1, -3)$ and $(x_2, y_2) = (3, 7)$

   Midpoint = $\left(\frac{1+3}{2}, \frac{-3+7}{2}\right) \rightarrow \left(\frac{4}{2}, \frac{4}{2}\right) \rightarrow M(2, 2)$

2) Find the midpoint of the line segment with the given endpoints. $(-4, 5), (8, -7)$

   Solution: Midpoint = $\left(\frac{x_1+x_2}{2}, \frac{y_1+y_2}{2}\right) \rightarrow (x_1, y_1) = (-4, 5)$ and $(x_2, y_2) = (8, -7)$

   Midpoint = $\left(\frac{-4+8}{2}, \frac{5-7}{2}\right) \rightarrow \left(\frac{4}{2}, \frac{-2}{2}\right) \rightarrow M(2, -1)$

3) Find the midpoint of the line segment with the given endpoints. $(5, -2), (1, 10)$

   Solution: Midpoint = $\left(\frac{x_1+x_2}{2}, \frac{y_1+y_2}{2}\right) \rightarrow (x_1, y_1) = (5, -2)$ and $(x_2, y_2) = (1, 10)$

   Midpoint = $\left(\frac{5+1}{2}, \frac{-2+10}{2}\right) \rightarrow \left(\frac{6}{2}, \frac{8}{2}\right) \rightarrow M(3, 4)$

4) Find the midpoint of the line segment with the given endpoints. $(2, 3), (12, -9)$

   Solution: Midpoint = $\left(\frac{x_1+x_2}{2}, \frac{y_1+y_2}{2}\right) \rightarrow (x_1, y_1) = (2, 3)$ and $(x_2, y_2) = (12, -3)$

   Midpoint = $\left(\frac{2+12}{2}, \frac{3-9}{2}\right) \rightarrow \left(\frac{14}{2}, \frac{-6}{2}\right) \rightarrow M(7, -3)$

Finding Distance of Two Points

✓ Use following formula to find the distance of two points with the coordinates $A\ (x_1, y_1)$ and $B\ (x_2, y_2)$:

$$d = \sqrt{(x_2 - x_1)^2 + (y_2 - y_1)^2}$$

Examples:

1) Find the distance between $(4, 6)$ and $(1, 2)$.

   Solution: *Use distance of two points formula:* $d = \sqrt{(x_2 - x_1)^2 + (y_2 - y_1)^2}$
   $(x_1, y_1) = (4, 6)$ and $(x_2, y_2) = (1, 2)$. Then: $d = \sqrt{(x_2 - x_1)^2 + (y_2 - y_1)^2} \rightarrow$
   $d = \sqrt{(1 - (4))^2 + (2 - 6)^2} = \sqrt{(-3)^2 + (-4)^2} = \sqrt{9 + 16} = \sqrt{25} = 5 \rightarrow d = 5$

2) Find the distance of two points $(-6, -10)$ and $(-2, -10)$.

   Solution: *Use distance of two points formula:* $d = \sqrt{(x_2 - x_1)^2 + (y_2 - y_1)^2}$
   $(x_1, y_1) = (-6, -10)$, and $(x_2, y_2) = (-2, -10)$
   Then: $d = \sqrt{(x_2 - x_1)^2 + (y_2 - y_1)^2} \rightarrow d = \sqrt{(-2 - (-6))^2 + (-10 - (-10))^2} =$
   $\sqrt{(4)^2 + (0)^2} = \sqrt{16 + 0} = \sqrt{16} = 4$. Then: $d = 4$

3) Find the distance between $(-6, 5)$ and $(-2, 2)$.

   Solution: *Use distance of two points formula:* $d = \sqrt{(x_2 - x_1)^2 + (y_2 - y_1)^2}$
   $(x_1, y_1) = (-6, 5)$ and $(x_2, y_2) = (-2, 2)$. Then: $d = \sqrt{(x_2 - x_1)^2 + (y_2 - y_1)^2} \rightarrow$
   $d = \sqrt{(-2 - (-6))^2 + (2 - 5)^2} = \sqrt{(4)^2 + (-3)^2} = \sqrt{16 + 9} = \sqrt{25} = 5$

Chapter 8: Practices

✍ *Find the slope of each line.*

1) $y = x - 1$
2) $y = -2x + 5$
3) $y = 2x - 1$
4) Line through $(-2, 4)$ and $(6, 0)$
5) Line through $(-3, 5)$ and $(-2, 8)$
6) Line through $(-3, -1)$ and $(0, -4)$

✍ *Sketch the graph of each line. (Using Slope–Intercept Form)*

7) $y = x + 2$

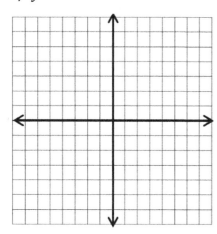

8) $y = 2x - 3$

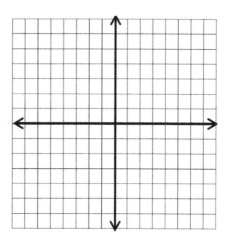

✍ *Solve.*

9) What is the equation of a line with slope 4 and intercept 12? _____

10) What is the equation of a line with slope 4 and passes through point $(4, 2)$?
_____

11) What is the equation of a line with slope -2 and passes through point $(-2, 4)$?
_____

12) The slope of a line is -3 and it passes through point $(-1, 5)$. What is the equation of the line? _____

13) The slope of a line is 3 and it passes through point $(-1, 4)$. What is the equation of the line? _____

✏️ *Sketch the graph of each linear inequality. (Graphing Linear Inequalities)*

15) $y > 3x - 1$

16) $y < -x + 4$

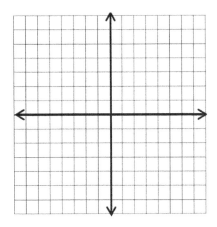
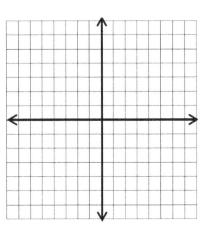

✏️ *Find the midpoint of the line segment with the given endpoints.*

17) $(2, 5), (-2, 7)$
18) $(-3, 6), (5, 2)$
19) $(9, -7), (-6, 5)$

20) $(14, -8), (-4, -2)$
21) $(-11, 2), (3, 8)$
22) $(7, 10), (1, -6)$

✏️ *Find the distance between each pair of points.*

23) $(3, 10), (-2, -2)$
24) $(1, 6), (4, 10)$
25) $(-2, -1), (-8, 7)$

26) $(8, -2), (5, -6)$
27) $(4, -3), (-5, 9)$
28) $(0, 6), (3, 2)$

Answers – Chapter 8

Find the slope of the line through each pair of points.

1) 1
2) −2
3) 2
4) $-\frac{1}{2}$
5) 3
6) −1

Sketch the graph of each line. (Using Slope–Intercept Form)

7) $y = x + 2$

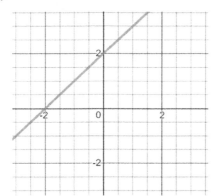

8) $y = 2x - 3$

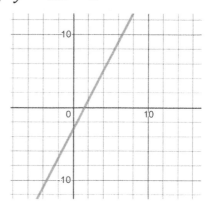

Write the equation of the line through the given points.

9) $y = 4x + 12$
10) $y = 4x - 14$
11) $y = -2x + 8$
12) $y = -3x + 2$
13) $y = 3x + 7$

Sketch the graph of each linear inequality. (Graphing Linear Inequalities)

15) $y > 3x - 1$

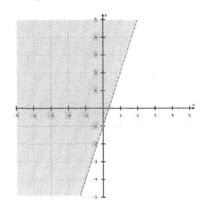

16) $y < -x + 4$

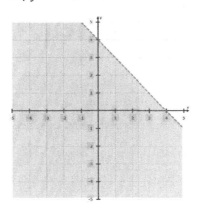

Find the midpoint of the line segment with the given endpoints.

17) $(0, 6)$
18) $(1, 4)$
19) $(1.5, -1)$

20) $(5, -5)$
21) $(-4, 5)$
22) $(4, 2)$

Find the distance between each pair of points.

23) 13
24) 5
25) 10

26) 5
27) 15
28) 5

Chapter 9:

Exponents and Variables

Math Topics that you'll learn in this Chapter:

- ✓ Multiplication Property of Exponents
- ✓ Division Property of Exponents
- ✓ Powers of Products and Quotients
- ✓ Zero and Negative Exponents
- ✓ Negative Exponents and Negative Bases
- ✓ Scientific Notation
- ✓ Radicals

Multiplication Property of Exponents

- Exponents are shorthand for repeated multiplication of the same number by itself. For example, instead of 2×2, we can write 2^2. For $3 \times 3 \times 3 \times 3$, we can write 3^4

- In algebra, a variable is a letter used to stand for a number. The most common letters are: $x, y, z, a, b, c, m,$ and n.

- Exponent's rules: $x^a \times x^b = x^{a+b}$, $\dfrac{x^a}{x^b} = x^{a-b}$

$$(x^a)^b = x^{a \times b} \qquad (xy)^a = x^a \times y^a \qquad \left(\dfrac{a}{b}\right)^c = \dfrac{a^c}{b^c}$$

Examples:

1) *Multiply.* $4x^3 \times 2x^2$

   Solution: Use Exponent's rules: $x^a \times x^b = x^{a+b} \to x^3 \times x^2 = x^{3+2} = x^5$

   Then: $4x^3 \times 2x^2 = 8x^5$

2) *Simplify.* $(x^3 y^5)^2$

   Solution: Use Exponent's rules: $(x^a)^b = x^{a \times b}$. Then: $(x^3 y^5)^2 = x^{3 \times 2} y^{5 \times 2} = x^6 y^{10}$

3) *Multiply.* $-2x^5 \times 7x^3$

   Solution: Use Exponent's rules: $x^a \times x^b = x^{a+b} \to x^5 \times x^3 = x^{5+3} = x^8$

   Then: $-2x^5 \times 7x^3 = -14x^8$

4) *Simplify.* $(x^2 y^4)^3$

   Solution: Use Exponent's rules: $(x^a)^b = x^{a \times b}$. Then: $(x^2 y^4)^3 = x^{2 \times 3} y^{4 \times 3} = x^6 y^{12}$

Division Property of Exponents

- ✓ Exponents are shorthand for repeated multiplication of the same number by itself. For example, instead of 3×3, we can write 3^2. For $2 \times 2 \times 2$, we can write 2^3

- ✓ For division of exponents use following formulas:

$$\frac{x^a}{x^b} = x^{a-b}, x \neq 0, \qquad \frac{x^a}{x^b} = \frac{1}{x^{b-a}}, x \neq 0, \qquad \frac{1}{x^b} = x^{-b}$$

Examples:

1) Simplify. $\frac{12x^2y}{4xy^3} =$

   Solution: First cancel the common factor: $4 \to \frac{12x^2y}{4xy^3} = \frac{3x^2y}{xy^3}$

   Use Exponent's rules: $\frac{x^a}{x^b} = x^{a-b} \to \frac{x^2}{x} = x^{2-1} = x$ and $\frac{y}{y^3} = \frac{1}{y^{3-1}} = \frac{1}{y^2}$

   Then: $\frac{12x^2y}{4xy^3} = \frac{3x}{y^2}$

2) Simplify. $\frac{18x^6}{2x^3} =$

   Solution: Use Exponent's rules: $\frac{x^a}{x^b} = x^{b-a} \to \frac{x^6}{x^3} = x^{6-3} = x^3$

   Then: $\frac{18x^6}{2x^3} = 9x^3$

3) Simplify. $\frac{8x^3y}{40x^2y^3} =$

   Solution: First cancel the common factor: $8 \to \frac{8x^3y}{40x^2y^3} = \frac{x^3y}{5x^2y^3}$

   Use Exponent's rules: $\frac{x^a}{x^b} = x^{a-b} \to \frac{x^3}{x^2} = x^{3-2} = x$

   Then: $\frac{8x^3y}{40x^2y^3} = \frac{xy}{5y^3} \to$ now cancel the common factor: $y \to \frac{xy}{5y^3} = \frac{x}{5y^2}$

Powers of Products and Quotients

- Exponents are shorthand for repeated multiplication of the same number by itself. For example, instead of $2 \times 2 \times 2$, we can write 2^3. For $3 \times 3 \times 3 \times 3$, we can write 3^4
- For any nonzero numbers a and b and any integer x, $(ab)^x = a^x \times b^x$ and $(\frac{a}{b})^c = \frac{a^c}{b^c}$

Examples:

1) Simplify. $(6x^2y^4)^2$

   Solution: Use Exponent's rules: $(x^a)^b = x^{a \times b}$

   $(6x^2y^4)^2 = (6)^2(x^2)^2(y^4)^2 = 36x^{2 \times 2}y^{4 \times 2} = 36x^4y^8$

2) Simplify. $(\frac{5x}{2x^2})^2$

   Solution: First cancel the common factor: $x \rightarrow (\frac{5x}{2x^2})^2 = (\frac{5}{2x})^2$

   Use Exponent's rules: $(\frac{a}{b})^c = \frac{a^c}{b^c}$, Then: $(\frac{5}{2x})^2 = \frac{5^2}{(2x)^2} = \frac{25}{4x^2}$

3) Simplify. $(3x^5y^4)^2$

   Solution: Use Exponent's rules: $(x^a)^b = x^{a \times b}$

   $(3x^5y^4)^2 = (3)^2(x^5)^2(y^4)^2 = 9x^{5 \times 2}y^{4 \times 2} = 9x^{10}y^8$

4) Simplify. $(\frac{2x}{3x^2})^2$

   Solution: First cancel the common factor: $x \rightarrow (\frac{2x}{3x^2})^2 = (\frac{2}{3x})^2$

   Use Exponent's rules: $(\frac{a}{b})^c = \frac{a^c}{b^c}$, Then: $(\frac{2}{3x})^2 = \frac{2^2}{(3x)^2} = \frac{4}{9x^2}$

Zero and Negative Exponents

- Zero-Exponent Rule: $a^0 = 1$, this means that anything raised to the zero power is 1. For example: $(5xy)^0 = 1$
- A negative exponent simply means that the base is on the wrong side of the fraction line, so you need to flip the base to the other side. For instance, "x^{-2}" (pronounced as "ecks to the minus two") just means "x^2" but underneath, as in $\frac{1}{x^2}$.

Examples:

1) Evaluate. $\left(\frac{2}{3}\right)^{-2} =$

   Solution: Use negative exponent's rule: $\left(\frac{x^a}{x^b}\right)^{-2} = \left(\frac{x^b}{x^a}\right)^{2} \to \left(\frac{2}{3}\right)^{-2} = \left(\frac{3}{2}\right)^{2} =$

   Then: $\left(\frac{3}{2}\right)^{2} = \frac{3^2}{2^2} = \frac{9}{4}$

2) Evaluate. $\left(\frac{4}{5}\right)^{-3} =$

   Solution: Use negative exponent's rule: $\left(\frac{x^a}{x^b}\right)^{-2} = \left(\frac{x^b}{x^a}\right)^{2} \to \left(\frac{4}{5}\right)^{-3} = \left(\frac{5}{4}\right)^{3} =$

   Then: $\left(\frac{5}{4}\right)^{3} = \frac{5^3}{4^3} = \frac{125}{64}$

3) Evaluate. $\left(\frac{x}{y}\right)^{0} =$

   Solution: Use zero-exponent Rule: $a^0 = 1$
   Then: $\left(\frac{x}{y}\right)^{0} = 1$

4) Evaluate. $\left(\frac{5}{6}\right)^{-1} =$

   Solution: Use negative exponent's rule: $\left(\frac{x^a}{x^b}\right)^{-2} = \left(\frac{x^b}{x^a}\right)^{2} \to \left(\frac{5}{6}\right)^{-1} = \left(\frac{6}{5}\right)^{1} = \frac{6}{5}$

Negative Exponents and Negative Bases

- A negative exponent is the reciprocal of that number with a positive exponent. $(3)^{-2} = \frac{1}{3^2}$
- To simplify a negative exponent, make the power positive!
- The parenthesis is important! -5^{-2} is not the same as $(-5)^{-2}$

$$-5^{-2} = -\frac{1}{5^2} \text{ and } (-5)^{-2} = +\frac{1}{5^2}$$

Examples:

1) Simplify. $\left(\frac{5a}{6c}\right)^{-2} =$

   Solution: Use negative exponent's rule: $\left(\frac{x^a}{x^b}\right)^{-2} = \left(\frac{x^b}{x^a}\right)^2 \rightarrow \left(\frac{5a}{6c}\right)^{-2} = \left(\frac{6c}{5a}\right)^2$

   Now use exponent's rule: $\left(\frac{a}{b}\right)^c = \frac{a^c}{b^c} \rightarrow = \left(\frac{6c}{5a}\right)^2 = \frac{6^2 c^2}{5^2 a^2}$

   Then: $\frac{6^2 c^2}{5^2 a^2} = \frac{36c^2}{25a^2}$

2) Simplify. $\left(\frac{2x}{3yz}\right)^{-3} =$

   Solution: Use negative exponent's rule: $\left(\frac{x^a}{x^b}\right)^{-2} = \left(\frac{x^b}{x^a}\right)^2 \rightarrow \left(\frac{2x}{3yz}\right)^{-3} = \left(\frac{3yz}{2x}\right)^3$

   Now use exponent's rule: $\left(\frac{a}{b}\right)^c = \frac{a^c}{b^c} \rightarrow \left(\frac{3yz}{2x}\right)^3 = \frac{3^3 y^3 z^3}{2^3 x^3} = \frac{27 y^3 z^3}{8 x^3}$

3) Simplify. $\left(\frac{3a}{2c}\right)^{-2} =$

   Solution: Use negative exponent's rule: $\left(\frac{x^a}{x^b}\right)^{-2} = \left(\frac{x^b}{x^a}\right)^2 \rightarrow \left(\frac{3a}{2c}\right)^{-2} = \left(\frac{2c}{3a}\right)^2$

   Now use exponent's rule: $\left(\frac{a}{b}\right)^c = \frac{a^c}{b^c} \rightarrow = \left(\frac{2c}{3a}\right)^2 = \frac{2^2 c^2}{3^2 a^2}$

   Then: $\frac{2^2 c^2}{3^2 a^2} = \frac{4c^2}{9a^2}$

Scientific Notation

- Scientific notation is used to write very big or very small numbers in decimal form.
- In scientific notation all numbers are written in the form of: $m \times 10^n$, where m is greater than 1 and less than 10.
- To convert a number from scientific notation to standard form, move the decimal point to the left (if the exponent of ten is a negative number), or to the right (if the exponent is positive).

Examples:

1) *Write 0.00015 in scientific notation.*

   Solution: First, move the decimal point to the right so that you have a number that is between 1 and 10. That number is 1.5
   Now, determine how many places the decimal moved in step 1 by the power of 10. We moved the decimal point 4 digits to the right. Then: 10^{-4} → When the decimal moved to the right, the exponent is negative. Then: $0.00015 = 1.5 \times 10^{-4}$

2) *Write 9.5×10^{-5} in standard notation.*

   Solution: 10^{-5} → When the decimal moved to the right, the exponent is negative.
   Then: $9.5 \times 10^{-5} = 0.000095$

3) *Write 0.00012 in scientific notation.*

   Solution: First, move the decimal point to the right so that you have a number that is between 1 and 10. Then: $m = 1.2$
   Now, determine how many places the decimal moved in step 1 by the power of 10.
   10^{-4} → Then: $0.00012 = 1.2 \times 10^{-4}$

4) *Write 8.3×10^5 in standard notation.*

   Solution: 10^{-5} → The exponent is positive 5. Then, move the decimal point to the right five digits. (remember $8.3 = 8.30000$)
   Then: $8.3 \times 10^5 = 830000$

Radicals

- If n is a positive integer and x is a real number, then: $\sqrt[n]{x} = x^{\frac{1}{n}}$, $\sqrt[n]{xy} = x^{\frac{1}{n}} \times y^{\frac{1}{n}}$, $\sqrt[n]{\frac{x}{y}} = \frac{x^{\frac{1}{n}}}{y^{\frac{1}{n}}}$, and $\sqrt[n]{x} \times \sqrt[n]{y} = \sqrt[n]{xy}$

- A square root of x is a number r whose square is: $r^2 = x$ (r is a square root of x.

- To add and subtract radicals, we need to have the same values under the radical. For example: $\sqrt{3} + \sqrt{3} = 2\sqrt{3}$, $3\sqrt{5} - \sqrt{5} = 2\sqrt{5}$

Examples:

1) *Find the square root of $\sqrt{169}$.*

   Solution: First factor the number: $169 = 13^2$,

   Then: $\sqrt{169} = \sqrt{13^2}$

   Now use radical rule: $\sqrt[n]{a^n} = a$.

   Then: $\sqrt{169} = \sqrt{13^2} = 13$

2) *Evaluate. $\sqrt{9} \times \sqrt{25} =$*

   Solution: Find the values of $\sqrt{9}$ and $\sqrt{25}$.

   Then: $\sqrt{9} \times \sqrt{25} = 3 \times 5 = 15$

3) *Solve. $7\sqrt{2} + 4\sqrt{2}$.*

   Solution: Since we have the same values under the radical, we can add these two radicals: $7\sqrt{2} + 4\sqrt{2} = 11\sqrt{2}$

4) *Evaluate. $\sqrt{2} \times \sqrt{8} =$*

   Solution: Use this radical rule: $\sqrt[n]{x} \times \sqrt[n]{y} = \sqrt[n]{xy} \rightarrow \sqrt{2} \times \sqrt{8} = \sqrt{16}$

   The square root of 16 is 4. Then: $\sqrt{2} \times \sqrt{8} = \sqrt{16} = 4$

Chapter 9: Practices

✎ *Simplify and write the answer in exponential form.*

1) $3x^3 \times 5xy^2 =$
2) $4x^2y \times 6x^2y^2 =$
3) $8x^3y^2 \times 2x^2y^3 =$
4) $7xy^4 \times 3x^2y =$
5) $6x^4y^5 \times 8x^3y^2 =$
6) $5x^3y^3 \times 8x^3y^3 =$

✎ *Simplify. (Division Property of Exponents)*

7) $\dfrac{5^5 \times 5^3}{5^9 \times 5} =$
8) $\dfrac{8x}{24x^2} =$
9) $\dfrac{15x^4}{9x^3} =$
10) $\dfrac{36x^3}{54x^3y^2} =$
11) $\dfrac{14^{\ 3}}{49^{\ 4}y^4} =$
12) $\dfrac{120x^3y^5}{30x^2y^3} =$

✎ *Simplify. (Powers of Products and Quotients)*

13) $(8x^4y^6)^3 =$
14) $(3x^5y^4)^6 =$
15) $(5x \times 4xy^2)^2 =$
16) $\left(\dfrac{6x}{x^3}\right)^2 =$
17) $\left(\dfrac{2x^3y^5}{6x^4y^2}\right)^2 =$
18) $\left(\dfrac{42^{\ 4}y^6}{21x^3y^5}\right)^3 =$

✎ *Evaluate the following expressions. (Zero and Negative Exponents)*

19) $\left(\dfrac{2}{5}\right)^{-2} =$
20) $\left(\dfrac{1}{2}\right)^{-8} =$
21) $\left(\dfrac{2}{5}\right)^{-3} =$
22) $\left(\dfrac{3}{7}\right)^{-2} =$
23) $\left(\dfrac{5}{6}\right)^{-3} =$
24) $\left(\dfrac{4}{9}\right)^{-2} =$

✎ *Simplify. (Negative Exponents and Negative Bases)*

25) $16x^{-3}y^{-4} =$
26) $-9x^2y^{-3} =$
27) $12a^{-4}b^2 =$
28) $25a^3b^{-5}c^{-1} =$
29) $\dfrac{18}{x^2y^{-2}} =$
30) $\dfrac{21^{\ -2}b}{-14c^{-4}} =$

✎ *Write each number in scientific notation.*

31) $0.00615 =$

32) $0.000048 =$

33) $36,000 =$

34) $82,000,000 =$

✎ *Evaluate.*

35) $\sqrt{7} \times \sqrt{7} =$ _____

36) $\sqrt{36} - \sqrt{9} =$ _____

37) $\sqrt{25} + \sqrt{49} =$ _____

38) $\sqrt{16} \times \sqrt{64} =$ _____

39) $\sqrt{3} \times \sqrt{12} =$ _____

40) $2\sqrt{6} + 3\sqrt{6} =$ _____

Answers – Chapter 9

1) $15x^4y^2$
2) $24x^4y^3$
3) $16x^5y^5$
4) $21x^3y^5$
5) $48x^7y^7$
6) $40x^6y^6$
7) $\frac{1}{25}$
8) $\frac{1}{3x}$
9) $\frac{5}{3}x$
10) $\frac{2}{3y^3}$
11) $\frac{2}{7x^4y}$
12) $4xy^2$
13) $512x^{12}y^{18}$
14) $729x^{30}y^{24}$
15) $400x^4y^4$
16) $\frac{36}{x^4}$
17) $\frac{y^6}{9x^2}$
18) $8x^3y^3$
19) $\frac{25}{4}$
20) 256
21) $\frac{125}{8}$
22) $\frac{49}{9}$
23) $\frac{216}{125}$
24) $\frac{81}{16}$
25) $\frac{16}{x^3y^4}$
26) $-\frac{9x^2}{y^3}$
27) $\frac{12b^2}{a^4}$
28) $\frac{25a^3}{b^5c}$
29) $\frac{18y^3}{x^2}$
30) $-\frac{3bc^4}{2a^2}$
31) 6.15×10^{-3}
32) 4.8×10^{-5}
33) 3.6×10^4
34) 8.2×10^7
35) 7
36) 3
37) 12
38) 32
39) 6
40) $5\sqrt{6}$

Chapter 10:

Polynomials

Math Topics that you'll learn in this Chapter:

- ✓ Simplifying Polynomials
- ✓ Adding and Subtracting Polynomials
- ✓ Multiplying Monomials
- ✓ Multiplying and Dividing Monomials
- ✓ Multiplying a Polynomial and a Monomial
- ✓ Multiplying Binomials
- ✓ Factoring Trinomials

Simplifying Polynomials

- To simplify Polynomials, find "like" terms. (they have same variables with same power).
- Use "FOIL". (First-Out-In-Last) for binomials:
$$(x + a)(x + b) = x^2 + (b + a)x + ab$$
- Add or Subtract "like" terms using order of operation.

Examples:

1) Simplify this expression. $x(2x + 5) + 6x =$

   Solution: Use Distributive Property: $x(2x + 5) = 2x^2 + 5x$
   Now, combine like terms: $x(2x + 5) + 6x = 2x^2 + 5x + 6x = 2x^2 + 11x$

2) Simplify this expression. $(x + 2)(x + 3) =$

   Solution: First apply FOIL method: $(a + b)(c + d) = ac + ad + bc + bd$

   $(x + 2)(x + 3) = x^2 + 3x + 2x + 6$

   Now combine like terms: $x^2 + 3x + 2x + 6 = x^2 + 5x + 6$

3) Simplify this expression. $4x(2x - 3) + 6x^2 - 4x =$

   Solution: Use Distributive Property: $4x(2x - 3) = 8x^2 - 12x$
   Then: $4x(2x - 3) + 6x^2 - 4x = 8x^2 - 12x + 6x^2 - 4x$
   Now combine like terms: $8x^2 + 6x^2 = 14x^2$, and $-12x - 4x = -16x$
   The simplified form of the expression: $8x^2 - 12x + 6x^2 - 4x = 14x^2 - 16x$

Adding and Subtracting Polynomials

- Adding polynomials is just a matter of combining like terms, with some order of operations considerations thrown in.
- Be careful with the minus signs, and don't confuse addition and multiplication!
- For subtracting polynomials, sometimes you need to use the Distributive Property: $a(b + c) = ab + ac$, $a(b - c) = ab - ac$

Examples:

1) Simplify the expressions. $(x^3 - 3x^4) - (2x^4 - 5x^3) =$

   Solution: First use Distributive Property: $-(2x^4 - 5x^3) = -1(2x^4 - 5x^3) = -2x^4 + 5x^3$

   $\rightarrow (x^3 - 3x^4) - (2x^4 - 5x^3) = x^3 - 3x^4 - 2x^4 + 5x^3$

   Now combine like terms: $x^3 + 5x^3 = 6x^3$ and $-3x^4 - 2x^4 = -5x^4$

   Then: $(x^3 - 3x^4) - (2x^4 - 5x^3) = x^3 - 3x^4 - 2x^4 + 5x^3 = 6x^3 - 5x^4$

   Write the answer in standard form: $6x^3 - 5x^4 = -5x^4 + 6x^3$

2) Add expressions. $(2x^3 - 4) + (6x^3 - 2x^2) =$

   Solution: Remove parentheses: $(2x^3 - 4) + (6x^3 - 2x^2) = 2x^3 - 4 + 6x^3 - 2x^2$

   Now combine like terms: $2x^3 - 4 + 6x^3 - 2x^2 = 8x^3 - 2x^2 - 4$

3) Simplify the expressions. $(8x^2 - 3x^3) - (2x^2 + 5x^3) =$

   Solution: First use Distributive Property: $-(2x^2 + 5x^3) = -2x^2 - 5x^3 \rightarrow$

   $(8x^2 - 3x^3) - (2x^2 + 5x^3) = 8x^2 - 3x^3 - 2x^2 - 5x^3$

   Now combine like terms and write in standard form: $8x^2 - 3x^3 - 2x^2 - 5x^3 = -8x^3 + 6x^2$

Multiplying Monomials

- A monomial is a polynomial with just one term: Examples: $2x$ or $7y^2$.
- When you multiply monomials, first multiply the coefficients (a number placed before and multiplying the variable) and then multiply the variables using multiplication property of exponents.

$$x^a \times x^b = x^{a+b}$$

Examples:

1) *Multiply expressions.* $5xy^4z^2 \times 3x^2y^5z^3$

   Solution: Find same variables and use multiplication property of exponents: $x^a \times x^b = x^{a+b}$

   $x \times x^2 = x^{1+2} = x^3$, $y^4 \times y^5 = y^{4+5} = y^9$ and $z^2 \times z^3 = z^{2+3} = z^5$

   Then, multiply coefficients and variables: $5xy^4z^2 \times 3x^2y^5z^3 = 15x^3y^9z^5$

2) *Multiply expressions.* $-2a^5b^4 \times 8a^3b^4 =$

   Solution: Use multiplication property of exponents: $x^a \times x^b = x^{a+b}$

   $a^5 \times a^3 = a^{5+3} = a^8$ and $b^4 \times b^4 = b^{4+4} = b^8$

   Then: $-2a^5b^4 \times 8a^3b^4 = -16a^8b^8$

3) *Multiply.* $7xy^3z^5 \times 4x^2y^4z^3$

   Solution: Use multiplication property of exponents: $x^a \times x^b = x^{a+b}$

   $x \times x^2 = x^{1+2} = x^3$, $y^3 \times y^4 = y^{3+4} = y^7$ and $z^5 \times z^3 = z^{5+3} = z^8$

   Then: $7xy^3z^5 \times 4x^2y^5z^3 = 28x^3y^7z^8$

4) *Simplify.* $(5a^6b^3)(-9a^7b^2) =$

   Solution: Use multiplication property of exponents: $x^a \times x^b = x^{a+b}$

   $a^6 \times a^7 = a^{6+7} = a^{13}$ and $b^3 \times b^2 = b^{3+2} = b^5$

   Then: $(5a^6b^3) \times (-9a^6b^2) = -45a^{13}b^5$

Multiplying and Dividing Monomials

- When you divide or multiply two monomials you need to divide or multiply their coefficients and then divide or multiply their variables.
- In case of exponents with the same base, for Division, subtract their powers, for Multiplication, add their powers.
- Exponent's Multiplication and Division rules:

$$x^a \times x^b = x^{a+b}, \qquad \frac{x^a}{x^b} = x^{a-b}$$

Examples:

1) *Multiply expressions.* $(-5x^8)(4x^6) =$

   Solution: Use multiplication property of exponents: $x^a \times x^b = x^{a+b} \rightarrow x^8 \times x^6 = x^{14}$

   Then: $(-5x^5)(4x^4) = -20x^{14}$

2) *Divide expressions.* $\frac{14x^5y^4}{2xy^3} =$

   Solution: Use division property of exponents: $\frac{x^a}{x^b} = x^{a-b} \rightarrow \frac{x^5}{x} = x^{5-1} = x^4$ and $\frac{y^4}{y^3} = y$

   Then:  $= 7x^4y$

3) *Divide expressions.* $\frac{56a^8b^3}{8ab^3}$

   Solution: Use division property of exponents: $\frac{x^a}{x^b} = x^{a-b} \rightarrow \frac{a^8}{a} = a^{8-1} = a^7$ and $\frac{b^3}{b^3} = 1$

   Then: $\frac{56a^8b^3}{8ab^3} = 7a^7$

Multiplying a Polynomial and a Monomial

✓ When multiplying monomials, use the product rule for exponents.
$$x^a \times x^b = x^{a+b}$$

✓ When multiplying a monomial by a polynomial, use the distributive property.
$$a \times (b + c) = a \times b + a \times c = ab + ac$$
$$a \times (b - c) = a \times b - a \times c = ab - ac$$

Examples:

1) *Multiply expressions.* $5x(3x - 2)$

   Solution: Use Distributive Property: $5x(3x - 2) = 5x \times 3x - 5x \times (-2) = 15x^2 - 10x$

2) *Multiply expressions.* $x(2x^2 + 3y^2)$

   Solution: Use Distributive Property: $x(2x^2 + 3y^2) = x \times 2x^2 + x \times 3y^2 = 2x^3 + 3xy^2$

3) *Multiply.* $-4x(-5x^2 + 3x - 6)$

   Solution: Use Distributive Property:

   $-4x(-5x^2 + 3x - 6) = (-4x)(-5x^2) + (-4x) \times (3x) + (-4x) \times (-6) =$

   Now, simplify: $(-4x)(-5x^2) + (-4x) \times (3x) + (-4x) \times (-6) = 20x^3 - 12x^2 + 24x$

Multiplying Binomials

☑ A binomial is a polynomial that is the sum or the difference of two terms, each of which is a monomial.

☑ To multiply two binomials, use "FOIL" method. (First-Out-In-Last)

$$(x + a)(x + b) = x \times x + x \times b + a \times x + a \times b = x^2 + bx + ax + ab$$

Examples:

1) *Multiply Binomials.* $(x + 2)(x - 4) =$

    Solution: Use "FOIL". (First–Out–In–Last): $(x + 2)(x - 4) = x^2 - 4x + 2x - 8$

    Then combine like terms: $x^2 - 4x + 2x - 8 = x^2 - 2x - 8$

2) *Multiply.* $(x - 5)(x - 2) =$

    Solution: Use "FOIL". (First–Out–In–Last): $(x - 5)(x - 2) = x^2 - 2x - 5x + 10$

    Then simplify: $x^2 - 2x - 5x + 10 = x^2 - 7x + 10$

3) *Multiply.* $(x - 3)(x + 6) =$

    Solution: Use "FOIL". (First–Out–In–Last): $(x - 3)(x + 6) = x^2 + 6x - 3x - 18$

    Then simplify: $x^2 + 6x - 3x - 18 = x^2 + 3x - 18$

4) *Multiply Binomials.* $(x + 8)(x + 4) =$

    Solution: Use "FOIL". (First–Out–In–Last): $(x + 8)(x + 4) = x^2 + 4x + 8x + 32$

    Then combine like terms: $x^2 + 4x + 8x + 32 = x^2 + 12x + 32$

Factoring Trinomials

To factor trinomials, you can use following methods:

☑ "FOIL": $(x + a)(x + b) = x^2 + (b + a)x + ab$

☑ "Difference of Squares":
$$a^2 - b^2 = (a + b)(a - b)$$
$$a^2 + 2ab + b^2 = (a + b)(a + b)$$
$$a^2 - 2ab + b^2 = (a - b)(a - b)$$

☑ "Reverse FOIL": $x^2 + (b + a)x + ab = (x + a)(x + b)$

Examples:

1) **Factor this trinomial.** $x^2 - 2x - 8$

   Solution: Break the expression into groups. You need to find two numbers that their product is -8 and their sum is -2. (remember "Reverse FOIL": $x^2 + (b + a)x + ab = (x + a)(x + b)$). Those two numbers are 2 and -4. Then:
   $x^2 - 2x - 8 = (x^2 + 2x) + (-4x - 8)$
   Now factor out x from $x^2 + 2x$: $x(x + 2)$, and factor out -4 from $-4x - 8$: $-4(x + 2)$
   Then: $(x^2 + 2x) + (-4x - 8) = x(x + 2) - 4(x + 2)$
   Now factor out like term: $(x + 2)$. Then: $(x + 2)(x - 4)$

2) **Factor this trinomial.** $x^2 - 2x - 24$

   Solution: Break the expression into groups: $(x^2 + 4x) + (-6x - 24)$
   Now factor out x from $x^2 + 4x$: $x(x + 4)$, and factor out -6 from $-6x - 24$: $-6(x + 4)$
   Then: $(x + 4) - 6(x + 4)$, now factor out like term:
   $(x = 4) \rightarrow x(x + 4) - 6(x + 4) = (x + 4)(x - 6)$

Chapter 10: Practices

✎ *Simplify each polynomial.*

1) $4(3x + 5) =$

2) $8(6x - 1) =$

3) $2x(7x + 4) + 2x =$

4) $5x(6x + 2) - 4x =$

5) $7x(3x - 6) + x^2 - 2 =$

6) $3x^2 - 5 - 9x(4x + 1) =$

✎ *Add or subtract polynomials.*

7) $(6x^2 + 8) + (2x^2 - 4) =$

8) $(x^2 - 5x) - (3x^2 + 2x) =$

9) $(9x^3 - 4x^2) + (x^3 - 3x^2) =$

10) $(7x^3 - 3x) - (6x^3 - 5x) =$

11) $(12x^3 + x^2) + (2x^2 - 10) =$

12) $(4x^3 - 15) - (3x^3 - 7x^2) =$

✎ *Simplify each expression. (Multiplying Monomials)*

13) $5x^3 \times 6x^4 =$

14) $-4a^2b \times 3ab^2 =$

15) $(-7x^2yz) \times (-3xy^3z^2) =$

16) $9u^4t^2 \times (-4ut) =$

17) $12x^3z \times 3xy^2 =$

18) $-8a^2bc \times a^3b^2 =$

✎ *Simplify each expression. (Multiplying and Dividing Monomials)*

19) $(4x^3y^4)(12x^5y^2) =$

20) $(5x^4y^2)(8x^6y^5) =$

21) $(16x^7y^9)(3x^6y^4) =$

22) $\frac{36x^5y^3}{9x^2y} =$

23) $\frac{98x^{12}y^{10}}{7x^9y^7} =$

24) $\frac{225x^9y^{13}}{15\ ^6y^9} =$

✍ Find each product. (Multiplying a Polynomial and a Monomial)

25) $4x(6x - y) =$

26) $7x(3x + 5y) =$

27) $5x(x - 8y) =$

28) $x(3x^2 + 2x - 6) =$

29) $5x(-x^2 + 7x + 4) =$

30) $6x(6x^2 - 3x - 12) =$

✍ Find each product. (Multiplying Binomials)

31) $(x - 3)(x + 3) =$

32) $(x - 5)(x - 4) =$

33) $(x + 6)(x + 3) =$

34) $(x - 7)(x + 8) =$

35) $(x + 2)(x - 9) =$

36) $(x - 15)(x + 3) =$

✍ Factor each trinomial.

37) $x^2 + 4x - 12 =$

38) $x^2 + x - 20 =$

39) $x^2 + 3x - 108 =$

40) $x^2 + 12x + 32 =$

41) $x^2 - 14x + 48 =$

42) $x^2 + 2x - 35 =$

Answers – Chapter 10

1) $12x + 20$
2) $48x - 8$
3) $14x^2 + 10x$

4) $30x^2 + 6x$
5) $22x^2 - 42x - 2$
6) $-33x^2 - 9x - 5$

7) $8x^2 + 4$
8) $-2x^2 - 7x$
9) $10x^3 - 7x^2$

10) $x^3 + 2x$
11) $12x^3 + 3x^2 - 10$
12) $x^3 + 7x^2 - 15$

13) $30x^7$
14) $-12a^3b^3$
15) $21x^3y^4z^3$

16) $-36u^5t^3$
17) $36x^4y^2z$
18) $-8a^5b^3c$

19) $48x^8y^6$
20) $40x^{10}y^7$
21) $48x^{13}y^{13}$
22) $4x^3y^2$

23) $14x^3y^3$

24) $15x^3y^4$

25) $24x^2 - 4xy$
26) $21x^2 + 35xy$
27) $5x^2 - 40xy$

28) $3x^3 + 2x^2 - 6x$
29) $-5x^3 + 35x^2 + 20x$
30) $36x^3 - 18x^2 - 72x$

31) $x^2 - 9$
32) $x^2 - 9x + 20$
33) $x^2 + 9x + 18$

34) $x^2 + x - 56$
35) $x^2 - 7x - 18$
36) $x^2 - 12x - 45$

37) $(x - 2)(x + 6)$
38) $(x + 5)(x - 4)$
39) $(x + 12)(x - 9)$

40) $(x + 8)(x + 4)$
41) $(x - 6)(x - 8)$
42) $(x + 7)(x - 5)$

Chapter 11:

Geometry and Solid Figures

Math Topics that you'll learn in this Chapter:

- ✓ The Pythagorean Theorem
- ✓ Triangles
- ✓ Polygons
- ✓ Circles
- ✓ Trapezoids
- ✓ Cubes
- ✓ Rectangle Prisms
- ✓ Cylinder

The Pythagorean Theorem

☑ You can use the Pythagorean Theorem to find a missing side in a right triangle.

☑ In any right triangle: $a^2 + b^2 = c^2$

Examples:

1) Right triangle ABC (not shown) has two legs of lengths 6 cm (AB) and 8 cm (AC). What is the length of the hypotenuse of the triangle (side BC)?

   Solution: Use Pythagorean Theorem: $a^2 + b^2 = c^2$, $a = 6$, and $b = 8$

   Then: $a^2 + b^2 = c^2 \rightarrow 6^2 + 8^2 = c^2 \rightarrow 36 + 64 = c^2 \rightarrow 100 = c^2 \rightarrow c = \sqrt{100} = 10$

   The length of the hypotenuse is 10 cm.

2) Find the hypotenuse of the following triangle.

   Solution: Use Pythagorean Theorem: $a^2 + b^2 = c^2$

   Then: $a^2 + b^2 = c^2 \rightarrow 12^2 + 5^2 = c^2 \rightarrow 144 + 25 = c^2$

   $c^2 = 169 \rightarrow c = \sqrt{169} = 13$

   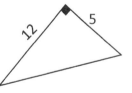

3) Find the length of the missing side in the following triangle.

   Solution: Use Pythagorean Theorem: $a^2 + b^2 = c^2$

   Then: $a^2 + b^2 = c^2 \rightarrow 3^2 + b^2 = 5^2 \rightarrow 9 + b^2 = 25 \rightarrow$

   $b^2 = 25 - 9 \rightarrow b^2 = 16 \rightarrow b = \sqrt{16} = 4$

   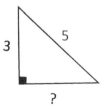

Triangles

✓ In any triangle the sum of all angles is 180 degrees.

✓ Area of a triangle $= \frac{1}{2}(base \times height)$

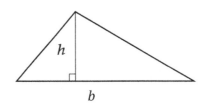

Examples:

What is the area of following triangles?

1)

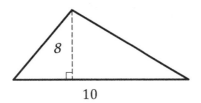

Solution:
Use the area formula: Area $= \frac{1}{2}(base \times height)$
$base = 10$ and $height = 8$
Area $= \frac{1}{2}(10 \times 8) = \frac{1}{2}(80) = 40$

2)

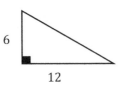

Solution:
Use the area formula: Area $= \frac{1}{2}(base \times height)$
$base = 12$ and $height = 6$
Area $= \frac{1}{2}(12 \times 6) = \frac{72}{2} = 36$

3) What is the missing angle in the following triangle?

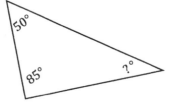

Solution:
In any triangle the sum of all angles is 180 degrees.
Let x be the missing angle. Then: $50 + 85 + x = 180$
→ $135 + x = 180$ → $x = 180 - 135 = 45$
The missing angle is 45 degrees.

Polygons

☑ Perimeter of a square
= 4 × side = 4s

☑ Perimeter of a rectangle
= 2(width + length)

☑ Perimeter of trapezoid
= a + b + c + d

☑ Perimeter of a regular hexagon = 6a

☑ Perimeter of a parallelogram = 2(l + w)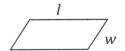

Examples:

1) Find the perimeter of following regular hexagon.

   Solution: Since the hexagon is regular, all sides are equal.

   Then: Perimeter of Hexagon = 6 × (one side)

   Perimeter of Hexagon = 6 × (one side) = 6 × 4 = 24 m

2) Find the perimeter of following trapezoid.

   Solution: Perimeter of a trapezoid = a + b + c + d

   Perimeter of the trapezoid = 5 + 6 + 6 + 8 = 25 ft

Circles

- In a circle, variable r is usually used for the radius and d for diameter.
- Area of a circle $= \pi r^2$ (π is about 3.14)
- Circumference of a circle $= 2\pi r$

Examples:

1) Find the area of the following circle.

    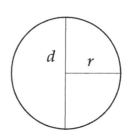

    Solution:

    Use area formula: $Area = \pi r^2$

    $r = 8\ in \rightarrow Area = \pi(8)^2 = 64\pi,\ \pi = 3.14$

    Then: $Area = 64 \times 3.14 = 200.96\ in^2$

2) Find the Circumference of the following circle.

    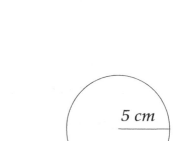

    Solution:

    Use Circumference formula: $Circumference = 2\pi r$

    $r = 5\ cm \rightarrow Circumference = 2\pi(5) = 10\pi$

    $\pi = 3.14$ **Then:** $Circumference = 10 \times 3.14 = 31.4\ cm$

3) Find the area of the circle.

    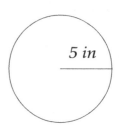

    Solution:

    Use area formula: $Area = \pi r^2$,

    $r = 5\ in$ then: $Area = \pi(5)^2 = 25\pi,\ \pi = 3.14$

    Then: $Area = 25 \times 3.14 = 78.5$

Trapezoids

✓ A quadrilateral with at least one pair of parallel sides is a trapezoid.

✓ Area of a trapezoid = $\frac{1}{2}h(b_1 + b_2)$

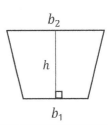

Examples:

1) Calculate the area of the following trapezoid.

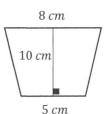

Solution:

Use area formula: $A = \frac{1}{2}h(b_1 + b_2)$

$b_1 = 5\ cm$, $b_2 = 8\ cm$ and $h = 10\ cm$

Then: $A = \frac{1}{2}(10)(8 + 5) = 5(13) = 65\ cm^2$

2) Calculate the area of the following trapezoid.

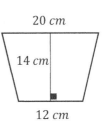

Solution:

Use area formula: $A = \frac{1}{2}h(b_1 + b_2)$

$b_1 = 12\ cm$, $b_2 = 20\ cm$ and $h = 14\ cm$

Then: $A = \frac{1}{2}(14)(12 + 20) = 7(32) = 224\ cm^2$

Cubes

- A cube is a three-dimensional solid object bounded by six square sides.
- Volume is the measure of the amount of space inside of a solid figure, like a cube, ball, cylinder or pyramid.
- Volume of a cube = $(one\ side)^3$
- surface area of a cube = $6 \times (one\ side)^2$

Examples:

1) Find the volume and surface area of the following cube.

   Solution: Use volume formula: $volume = (one\ side)^3$

   Then: $volume = (one\ side)^3 = (2)^3 = 8\ cm^3$

   Use surface area formula: $surface\ area\ of\ cube: 6(one\ side)^2 = 6(2)^2 = 6(4) = 24\ cm^2$

2) Find the volume and surface area of the following cube.

   Solution: Use volume formula: $volume = (one\ side)^3$

   Then: $volume = (one\ side)^3 = (5)^3 = 125\ cm^3$

   Use surface area formula:

   $surface\ area\ of\ cube: 6(one\ side)^2 = 6(5)^2 = 6(25) = 150\ cm^2$

   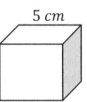

3) Find the volume and surface area of the following cube.

   Solution: Use volume formula: $volume = (one\ side)^3$

   Then: $volume = (one\ side)^3 = (7)^3 = 343\ m^3$

   Use surface area formula:

   $surface\ area\ of\ cube: 6(one\ side)^2 = 6(7)^2 = 6(49) = 294\ m^2$

   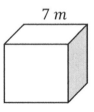

Rectangular Prisms

☑ A rectangular prism is a solid 3-dimensional object which has six rectangular faces.

☑ Volume of a Rectangular prism = **Length × Width × Height**

$Volume = l \times w \times h$

$Surface\ area = 2 \times (wh + lw + lh)$

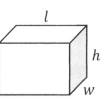

Examples:

1) Find the volume and surface area of the following rectangular prism.

    Solution:

    Use volume formula: $Volume = l \times w \times h$

    Then: $Volume = 8 \times 6 \times 10 = 480\ m^3$

    Use surface area formula: $Surface\ area = 2 \times (wh + lw + lh)$

    Then: $Surface\ area = 2 \times \big((6 \times 10) + (8 \times 6) + (8 \times 10)\big)$

    $\qquad = 2 \times (60 + 48 + 80) = 2 \times (188) = 376\ m^2$

2) Find the volume and surface area of rectangular prism.

    Solution:

    Use volume formula: $Volume = l \times w \times h$

    Then: $Volume = 10 \times 8 \times 12 = 960\ m^3$

    Use surface area formula: $Surface\ area = 2 \times (wh + lw + lh)$

    Then: $Surface\ area = 2 \times \big((8 \times 12) + (10 \times 8) + (10 \times 12)\big)$

    $\qquad = 2 \times (96 + 80 + 120) = 2 \times (296) = 592\ m^2$

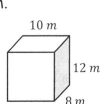

Cylinder

✓ A cylinder is a solid geometric figure with straight parallel sides and a circular or oval cross section.

✓ $Volume\ of\ a\ Cylinder = \pi(radius)^2 \times height,\ \pi \approx 3.14$

✓ $Surface\ area\ of\ a\ cylinder = 2\pi r^2 + 2\pi rh$

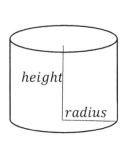

Examples:

1) Find the volume and Surface area of the follow Cylinder.

Solution:

Use volume formula: $Volume = \pi(radius)^2 \times height$
Then: $Volume = \pi(3)^2 \times 8 = 9\pi \times 8 = 72\pi$
$\pi = 3.14$ **then:** $Volume = 72\pi = 72 \times 3.14 = 226.08\ cm^3$
Use surface area formula: $Surface\ area = 2\pi r^2 + 2\pi rh$
Then: $2\pi(3)^2 + 2\pi(3)(8) = 2\pi(9) + 2\pi(24) = 18\pi + 48\pi = 66\pi$
$\pi = 3.14$ Then: $Surface\ area = 66 \times 3.14 = 207.24\ cm^2$

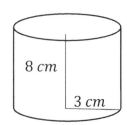

2) Find the volume and Surface area of the follow Cylinder.

Solution:

Use volume formula: $Volume = \pi(radius)^2 \times height$
Then: $Volume = \pi(2)^2 \times 6 = \pi 4 \times 6 = 24\pi$
$\pi = 3.14$ **then:** $Volume = 24\pi = 75.36\ cm^3$
Use surface area formula: $Surface\ area = 2\pi r^2 + 2\pi rh$
Then: $= 2\pi(2)^2 + 2\pi(2)(6) = 2\pi(4) + 2\pi(12) = 8\pi + 24\pi = 32\pi$
$\pi = 3.14$ **then:** $Surface\ area = 32 \times 3.14 = 100.48\ cm^2$

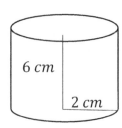

Chapter 11: Practices

✎ **Find the missing side?**

1)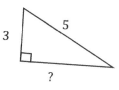
2)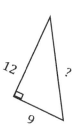
3)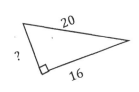
4)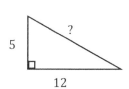

✎ **Find the measure of the unknown angle in each triangle.**

5)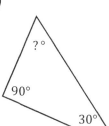
6)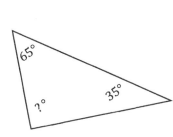
7)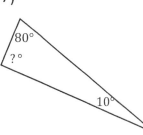
8)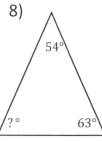

✎ **Find area of each triangle.**

9)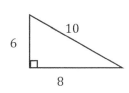
10)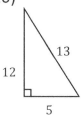
11)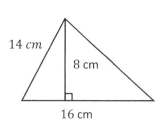
12)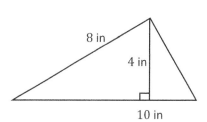

✎ **Find the perimeter of each shape.**

13)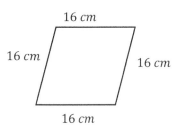
14)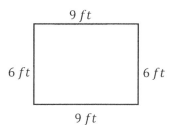
15)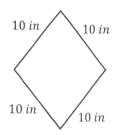
16)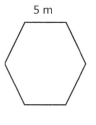

www.EffortlessMath.com

107

✍ **Complete the table below.** (π = 3.14)

17)

	Radius	Diameter	Circumference	Area
Circle 1	3 inches	6 inches	18.84 inches	28.26 square inches
Circle 2			43.96 meters	
Circle 3		8 ft		
Circle 4				78.5 square miles

✍ **Find the area of each trapezoid.**

18)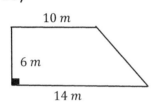

19)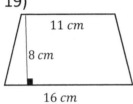

20)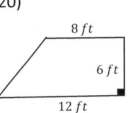

21)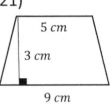

✍ **Find the volume of each cube.**

22)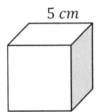

23)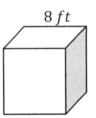

24)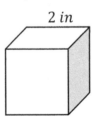

25)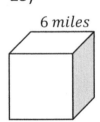

✍ **Find the volume of each Rectangular Prism.**

26)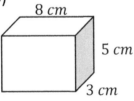

27)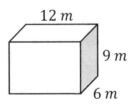

28)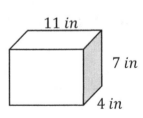

✏️ **Find the volume of each Cylinder. Round your answer to the nearest tenth.** ($\pi = 3.14$)

25) 4 cm, 14 cm

26) 6 m, 8 m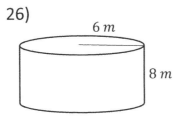

27) 8 cm, 10 cm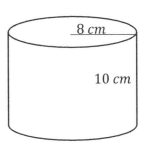

Answers – Chapter 11

1) 4
2) 15

3) 12
4) 13

5) 60°
6) 80°
7) 90°
8) 63°

9) 24 square unites
10) 30 square unites
11) 64 cm^2
12) 20 in^2

13) 64 cm
14) 30 ft
15) 40 in
16) 30 m

17)

	Radius	Diameter	Circumference	Area
Circle 1	3 inches	6 inches	18.84 inches	28.26 square inches
Circle 2	7 meters	14 meters	43.96 meters	153.86 square meters
Circle 3	4 ft	8 ft	25.12 ft	50.24 square ft
Circle 4	5 miles	10 miles	31.4 miles	78.5 square miles

18) 72 m^2
19) 108 cm^2
20) 60 ft^2
21) 21 cm^2

22) 125 cm^3
23) 512 ft^3
24) 8 in^3
25) 216 $miles^3$

26) 120 cm^3
27) 648 m^3
28) 308 in^3

29) 703.36 cm^3
30) 904.32 m^3
31) 2,009.6 cm^3

Chapter 12:

Statistics

Math Topics that you'll learn in this Chapter:

- ✓ Mean, Median, Mode, and Range of the Given Data

- ✓ Pie Graph

- ✓ Probability Problems

- ✓ Permutations and Combinations

Mean, Median, Mode, and Range of the Given Data

- Mean: $\dfrac{sum\ of\ the\ data}{total\ number\ of\ data\ entires}$

- Mode: the value in the list that appears most often

- Median: is the middle number of a group of numbers that have been arranged in order by size.

- Range: the difference of largest value and smallest value in the list

Examples:

1) What is the mode of these numbers? $4, 5, 7, 5, 7, 4, 0, 4$

   Solution: Mode: the value in the list that appears most often.
   Therefore, the mode is number 4. There are three number 4 in the data.

2) What is the median of these numbers? $5, 10, 14, 9, 16, 19, 6$

   Solution: Write the numbers in order: $5, 6, 9, 10, 14, 16, 19$

   Median is the number in the middle. Therefore, the median is 10.

3) What is the mean of these numbers? $8, 2, 8, 5, 3, 2, 4, 8$

   Solution: Mean: $\dfrac{sum\ of\ the\ data}{total\ number\ of\ data\ entires} = \dfrac{8+2+8+5+3+2+4+8}{8} = 5$

4) What is the range in this list? $4, 9, 13, 8, 15, 18, 5$

   Solution: Range is the difference of largest value and smallest value in the list. The largest value is 18 and the smallest value is 4. Then: $18 - 4 = 14$

Pie Graph

✓ A Pie Chart is a circle chart divided into sectors, each sector represents the relative size of each value.

✓ Pie charts represent a snapshot of how a group is broken down into smaller pieces.

Example:

A library has 820 books that include Mathematics, Physics, Chemistry, English and History. Use following graph to answer the questions.

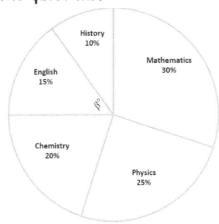

1) What is the number of Mathematics books?

   Solution: Number of total books = 820

   Percent of Mathematics books = 30% = 0.30

   Then, number of Mathematics books:
   $$0.30 \times 820 = 246$$

2) What is the number of History books?

   Solution: Number of total books = 820

   Percent of History books = 10% = 0.10

   Then: $0.10 \times 820 = 82$

3) What is the number of Chemistry books?

   Solution: Number of total books = 820

   Percent of Chemistry books = 20% = 0.20

   Then: $0.20 \times 820 = 164$

Probability Problems

- ✓ Probability is the likelihood of something happening in the future. It is expressed as a number between zero (can never happen) to 1 (will always happen).
- ✓ Probability can be expressed as a fraction, a decimal, or a percent.
- ✓ Probability formula: $Probability = \frac{number\ of\ desired\ outcomes}{number\ of\ total\ outcomes}$

Examples:

1) Anita's trick-or-treat bag contains 12 pieces of chocolate, 18 suckers, 18 pieces of gum, 24 pieces of licorice. If she randomly pulls a piece of candy from her bag, what is the probability of her pulling out a piece of sucker?

    Solution: $Probability = \frac{number\ of\ desired\ outcomes}{number\ of\ total\ outcomes}$

    Probability of ulling out a piece of sucker $= \frac{18}{12 + 18 + 18 + 24} = \frac{18}{72} = \frac{1}{4}$

2) A bag contains 20 balls: four green, five black, eight blue, a brown, a red and one white. If 19 balls are removed from the bag at random, what is the probability that a brown ball has been removed?

    Solution: If 19 balls are removed from the bag at random, there will be one ball in the bag. The probability of choosing a brown ball is 1 out of 20. Therefore, the probability of not choosing a brown ball is 19 out of 20 and the probability of having not a brown ball after removing 19 balls is the same.

Permutations and Combinations

- Factorials are products, indicated by an exclamation mark. For example, $4! = 4 \times 3 \times 2 \times 1$ (Remember that $0!$ is defined to be equal to 1.)
- Permutations: The number of ways to choose a sample of k elements from a set of n distinct objects where order does matter, and replacements are not allowed. For a permutation problem, use this formula:

$$_nP_k = \frac{n!}{(n-k)!}$$

- Combination: The number of ways to choose a sample of r elements from a set of n distinct objects where order does not matter, and replacements are not allowed. For a combination problem, use this formula:

$$_nC_r = \frac{n!}{r!\,(n-r)!}$$

Examples:

1) *How many ways can the first and second place be awarded to 8 people?*

    Solution: Since the order matters, (the first and second place are different!) we need to use permutation formula where n is 10 and k is 2. Then: $\frac{n!}{(n-k)!} = \frac{8!}{(8-2)!} = \frac{8!}{6!} = \frac{8 \times 7 \times 6!}{6!}$, remove 6! from both sides of the fraction. Then: $\frac{8 \times 7 \times 6!}{6!} = 8 \times 7 = 56$

2) *How many ways can we pick a team of 2 people from a group of 6?*

    Solution: Since the order doesn't matter, we need to use combination formula where n is 8 and r is 3. Then: $\frac{n!}{r!\,(n-r)!} = \frac{6!}{2!\,(6-2)!} = \frac{6!}{2!\,(4)!} = \frac{6 \times 5 \times 4!}{2!\,(4)!} = \frac{6 \times 5}{2 \times 1} = \frac{30}{2} = 15$

Chapter 12: Practices

✎ **Find the values of the Given Data.**

1) 6, 12, 1, 1, 5

   Mode: _____   Range: _____

   Mean: _____   Median: _____

2) 5, 8, 3, 7, 4, 3

   Mode: _____   Range: _____

   Mean: _____   Median: _____

✎ The circle graph below shows all Jason's expenses for last month. Jason spent $864 on his bills last month.

3) How much did Jason spend on his car last month? _____

4) How much did Jason spend for foods last month? _____

Jason's last month expenses

✎ *Solve.*

5) Bag A contains 8 red marbles and 6 green marbles. Bag B contains 5 black marbles and 10 orange marbles. What is the probability of selecting a green marble at random from bag A? What is the probability of selecting a black marble at random from Bag B? _____   _____

✎ *Solve.*

6) Susan is baking cookies. She uses sugar, flour, butter, and eggs. How many different orders of ingredients can she try? _____

7) Jason is planning for his vacation. He wants to go to museum, watch a movie, go to the beach, and play volleyball. How many different ways of ordering are there for him? _____

8) In how many ways a team of 10 basketball players can to choose a captain and co-captain? _____

9) How many ways can you give 3 balls to your 8 friends? _____

10) A professor is going to arrange her 8 students in a straight line. In how many ways can she do this? _____

Answers – Chapter 12

1) Mode: 1, Range: 11, Mean: 5, Median: 5

2) Mode: 3, Range: 5, Mean: 5, Median: 4.5

3) $640

4) $448

5) $\frac{3}{7}, \frac{1}{3}$

6) 24

7) 24

8) 90

9) 56

10) 40,320

Chapter 13:

Functions Operations

Math Topics that you'll learn in this Chapter:

- ✓ Function Notation

- ✓ Adding and Subtracting Functions

- ✓ Multiplying and Dividing Functions

- ✓ Composition of Functions

Function Notation and Evaluation

- ✓ Functions are mathematical operations that assign unique outputs to given inputs.
- ✓ Function notation is the way a function is written. It is meant to be a precise way of giving information about the function without a rather lengthy written explanation.
- ✓ The most popular function notation is $f(x)$ which is read "f of x". Any letter can be used to name a function. for example: $g(x), h(x)$, etc.
- ✓ To evaluate a function, plug in the input (the given value or expression) for the function's variable (place holder, x).

Examples:

1) Evaluate: $h(n) = n - 2$, find $h(2)$

   Solution: Substitute n with 4: Then: $h(n) = n - 2 \rightarrow h(2) = (2)^2 - 2 \rightarrow h(2) = 4 - 2 = 2$

2) Evaluate: $w(x) = 5x - 1$, find $w(3)$.

   Solution: Substitute x with 3: Then: $w(x) = 5x - 1 \rightarrow w(3) = 5(3) - 1 = 15 - 1 = 14$

3) Evaluate: $f(x) = x^2 - 2$, find $h(2)$.

   Solution: Substitute x with 2: Then: $f(x) = x^2 - 2 \rightarrow f(2) = (2)^2 - 2 = 4 - 2 = 2$

4) Evaluate: $p(x) = 2x^2 - 4$, find $p(3a)$.

   Solution: Substitute x with $3a$: Then: $p(x) = 2x^2 - 4 \rightarrow p(3a) = 2(3a)^2 - 4 \rightarrow$

   $$p(3a) = 2(9a^2) - 4 = 18a^2 - 4$$

Adding and Subtracting Functions

✓ Just like we can add and subtract numbers and expressions, we can add or subtract two functions and simplify or evaluate them. The result is a new function.

✓ For two functions $f(x)$ and $g(x)$, we can create two new functions:
$(f + g)(x) = f(x) + g(x)$ and $(f - g)(x) = f(x) - g(x)$

Examples:

1) $g(x) = a - 1, f(a) = a + 2$, Find: $(g + f)(a)$

   Solution: $(g + f)(a) = g(a) + f(a)$

   Then: $(g + f)(a) = (a - 1) + (a + 2) = 2a + 1$

2) $f(x) = 2x - 2, g(x) = x - 4$, Find: $(f - g)(x)$

   Solution: $(f - g)(x) = f(x) - g(x)$

   Then: $(f - g)(x) = (2x - 2) - (x - 4) = 2x - 2 - x + 4 = x + 2$

3) $g(x) = x^2 - 4, f(x) = 2x + 3$, Find: $(g + f)(x)$

   Solution: $(g + f)(x) = g(x) + f(x)$

   Then: $(g + f)(x) = (x^2 - 4) + (2x + 3) = x^2 + 2x - 1$

4) $f(x) = 2x^2 + 5, g(x) = 3x - 1$, Find: $(f - g)(5)$

   Solution: $(f - g)(x) = f(x) - g(x)$

   Then: $(f - g)(x) = (2x^2 + 5) - (3x - 1) = 2x^2 + 5 - 3x + 1 = 2x^2 - 3x + 6$

   Substitute x with 5: $(g - f)(5) = 2(5)^2 - 3(5) + 6 = 50 - 15 + 6 = 41$

Multiplying and Dividing Functions

✓ Just like we can multiply and divide numbers and expressions, we can multiply and divide two functions and simplify or evaluate them.

✓ For two functions $f(x)$ and $g(x)$, we can create two new functions:

$(f \cdot g)(x) = f(x) \cdot g(x)$ and $\left(\dfrac{f}{g}\right)(x) = \dfrac{f(x)}{g(x)}$

Examples:

1) $g(x) = x - 2, f(x) = x + 3$, Find: $(g \cdot f)(x)$
   Solution: $(g \cdot f)(x) = g(x) \cdot f(x) = (x - 2)(x + 3) = x^2 + 3x - 2x - 6$

2) $f(x) = x + 4, h(x) = x - 6$, Find: $\left(\dfrac{f}{h}\right)(x)$
   Solution: $\left(\dfrac{f}{h}\right)(x) = \dfrac{f(x)}{h(x)} = \dfrac{x+4}{x-6}$

3) $g(x) = x + 5, f(x) = x - 2$, Find: $(g \cdot f)(4)$
   Solution: $(g \cdot f)(x) = g(x) \cdot f(x) = (x + 5)(x - 2) = x^2 - 2x + 5x - 10 = x^2 + 3x - 10$
   Substitute x with 4: $(g \cdot f)(x) = (4)^2 + 3(4) - 10 = 16 + 12 - 10 = 18$

4) $f(x) = 2x + 3, h(x) = x + 8$, Find: $\left(\dfrac{f}{h}\right)(-1)$
   Solution: $\left(\dfrac{f}{h}\right)(x) = \dfrac{f(x)}{h(x)} = \dfrac{2x+3}{x+8}$
   Substitute x with -1: $\left(\dfrac{f}{h}\right)(x) = \dfrac{2x+3}{x+8} = \dfrac{2(-1)+3}{(-1)+8} = \dfrac{1}{7}$

Chapter 13: Practices

✎ *Evaluate each function.*

1) $g(n) = 6n - 3$, find $g(-2)$

2) $h(x) = -8x + 12$, find $h(3)$

3) $k(n) = 14 - 3n$, find $k(3)$

4) $g(x) = 4x - 4$, find $g(-2)$

5) $k(n) = 8n - 7$, find $k(4)$

6) $w(n) = -2n + 14$, find $w(5)$

✎ *Perform the indicated operation.*

7) $f(x) = x + 6$
   $g(x) = 3x + 3$
   Find $(f - g)(2)$

8) $g(x) = x - 3$
   $f(x) = -x - 4$
   Find $(g - f)(-2)$

9) $h(t) = 5t + 4$
   $g(t) = 2t + 2$
   Find $(h + g)(-1)$

10) $g(a) = 3a - 5$
    $f(a) = a^2 + 6$
    Find $(g + f)(3)$

11) $g(x) = 4x - 5$
    $h(x) = 6x^2 + 5$
    Find $(g - f)(-2)$

12) $h(x) = x^2 + 3$
    $g(x) = -4x + 1$
    Find $(h + g)(4)$

Perform the indicated operation.

13) $g(x) = x + 2$
    $f(x) = x + 3$
    Find $(g \cdot f)(4)$

    _____

14) $f(x) = 2x$
    $h(x) = -x + 6$
    Find $(f \cdot h)(-2)$

    _____

15) $g(a) = a + 2$
    $h(a) = 2a - 3$
    Find $(g \cdot h)(5)$

    _____

16) $f(x) = 2x + 4$
    $h(x) = 4x - 2$
    Find $\left(\dfrac{f}{h}\right)(2)$

    _____

17) $f(x) = a^2 - 2$
    $g(x) = -4 + 3a$
    Find $\left(\dfrac{f}{g}\right)(2)$

    _____

18) $g(a) = 4a + 6$
    $f(a) = 2a - 8$
    Find $\left(\dfrac{g}{f}\right)(3)$

    _____

Answers – Chapter 13

1) -15
2) -12
3) 5
4) -12
5) 25
6) 4
7) -1
8) -3
9) -1
10) 19

11) -42
12) 4
13) 42
14) -32
15) 49
16) $\frac{4}{3}$
17) 1
18) -9

ASVAB Test Review

The Armed Services Vocational Aptitude Battery (ASVAB) was introduced in 1968. Over 40 million examinees have taken the ASVAB since then.

According to official ASVAB website, the ASVAB is a multiple-aptitude battery that measures developed abilities and helps predict future academic and occupational success in the military. It is administered annually to more than one million military applicants, high school, and post-secondary students.

ASVAB scores are reported as percentiles between 1-99. An ASVAB percentile score indicates the percentage of examinees in a reference group that scored at or below that particular score. For example, ASVAB score of 90 indicates that the examinee scored as well as or better than 90% of the nationally-representative sample test takers. An ASVAB score of 60 indicates that the examinee scored as well as or better than 60% of the nationally-representative sample.

There are three types of ASVAB:

- The CAT-ASVAB (computer adaptive test)
- The MET-site ASVAB (paper and pencil (P&P)
- The Student ASVAB (paper and pencil (P&P)

The CAT-ASVAB is a computer adaptive test. It means that if the correct answer is chosen, the next question will be harder. If the answer given is incorrect, the next question will be easier. This also means that once an answer is selected on the CAT it cannot be changed.

The MET- site ASVAB and The Student ASVAB are paper and pencil (P&P) tests.

In this section, there are 2 complete Arithmetic Reasoning and Mathematics Knowledge ASVAB Tests. There is a complete test for CAT-ASVAB and another complete test for paper and pencil (P&P). Take these tests to see what score you'll be able to receive on a real ASVAB test.

Good luck!

Time to Test

Time to refine your Math skill with a practice test

In this section, there are two complete ASVAB Mathematics practice tests, one computer based (CAT-ASVAB) and one Paper and Pencil test. Take these tests to simulate the test day experience. After you've finished, score your tests using the answer keys.

Before You Start

- You'll need a pencil and a timer to take the test.
- For each question, there are four possible answers. Choose which one is best.
- It's okay to guess. There is no penalty for wrong answers.
- Use the answer sheet provided to record your answers.
- After you've finished the test, review the answer key to see where you went wrong.

Calculators are NOT permitted for the ASVAB Test

Good Luck!

Mathematics is like love; a simple idea, but it can get complicated.

ASVAB Math Practice Test 1 Answer Sheet

Remove (or photocopy) this answer sheet and use it to complete the practice test.

ASVAB Math Practice Test 1 (CAT-ASVAB) Answer Sheet

ASVAB Practice Test 1 — Arithmetic Reasoning

1. Ⓐ Ⓑ Ⓒ Ⓓ Ⓔ
2. Ⓐ Ⓑ Ⓒ Ⓓ Ⓔ
3. Ⓐ Ⓑ Ⓒ Ⓓ Ⓔ
4. Ⓐ Ⓑ Ⓒ Ⓓ Ⓔ
5. Ⓐ Ⓑ Ⓒ Ⓓ Ⓔ
6. Ⓐ Ⓑ Ⓒ Ⓓ Ⓔ
7. Ⓐ Ⓑ Ⓒ Ⓓ Ⓔ
8. Ⓐ Ⓑ Ⓒ Ⓓ Ⓔ
9. Ⓐ Ⓑ Ⓒ Ⓓ Ⓔ
10. Ⓐ Ⓑ Ⓒ Ⓓ Ⓔ
11. Ⓐ Ⓑ Ⓒ Ⓓ Ⓔ
12. Ⓐ Ⓑ Ⓒ Ⓓ Ⓔ
13. Ⓐ Ⓑ Ⓒ Ⓓ
14. Ⓐ Ⓑ Ⓒ Ⓓ
15. Ⓐ Ⓑ Ⓒ Ⓓ
16. Ⓐ Ⓑ Ⓒ Ⓓ
17.
18.
19.
20.

ASVAB Practice Test 1 — Mathematics Knowledge

1. Ⓐ Ⓑ Ⓒ Ⓓ Ⓔ
2. Ⓐ Ⓑ Ⓒ Ⓓ Ⓔ
3. Ⓐ Ⓑ Ⓒ Ⓓ Ⓔ
4. Ⓐ Ⓑ Ⓒ Ⓓ Ⓔ
5. Ⓐ Ⓑ Ⓒ Ⓓ Ⓔ
6. Ⓐ Ⓑ Ⓒ Ⓓ Ⓔ
7. Ⓐ Ⓑ Ⓒ Ⓓ Ⓔ
8. Ⓐ Ⓑ Ⓒ Ⓓ Ⓔ
9. Ⓐ Ⓑ Ⓒ Ⓓ Ⓔ
10. Ⓐ Ⓑ Ⓒ Ⓓ Ⓔ
11. Ⓐ Ⓑ Ⓒ Ⓓ Ⓔ
12. Ⓐ Ⓑ Ⓒ Ⓓ Ⓔ
13. Ⓐ Ⓑ Ⓒ Ⓓ Ⓔ
14. Ⓐ Ⓑ Ⓒ Ⓓ Ⓔ
15. Ⓐ Ⓑ Ⓒ Ⓓ Ⓔ
16. Ⓐ Ⓑ Ⓒ Ⓓ Ⓔ

ASVAB Math Practice Test 2 (Paper and Pencil) Answer Sheet

ASVAB Practice Test 2
Arithmetic Reasoning

1	Ⓐ Ⓑ Ⓒ Ⓓ Ⓔ	11 Ⓐ Ⓑ Ⓒ Ⓓ Ⓔ	21 Ⓐ Ⓑ Ⓒ Ⓓ Ⓔ
2	Ⓐ Ⓑ Ⓒ Ⓓ Ⓔ	12 Ⓐ Ⓑ Ⓒ Ⓓ Ⓔ	22 Ⓐ Ⓑ Ⓒ Ⓓ Ⓔ
3	Ⓐ Ⓑ Ⓒ Ⓓ Ⓔ	13 Ⓐ Ⓑ Ⓒ Ⓓ Ⓔ	23 Ⓐ Ⓑ Ⓒ Ⓓ Ⓔ
4	Ⓐ Ⓑ Ⓒ Ⓓ Ⓔ	14 Ⓐ Ⓑ Ⓒ Ⓓ Ⓔ	24 Ⓐ Ⓑ Ⓒ Ⓓ Ⓔ
5	Ⓐ Ⓑ Ⓒ Ⓓ Ⓔ	15 Ⓐ Ⓑ Ⓒ Ⓓ Ⓔ	25 Ⓐ Ⓑ Ⓒ Ⓓ Ⓔ
6	Ⓐ Ⓑ Ⓒ Ⓓ Ⓔ	16 Ⓐ Ⓑ Ⓒ Ⓓ Ⓔ	26 Ⓐ Ⓑ Ⓒ Ⓓ Ⓔ
7	Ⓐ Ⓑ Ⓒ Ⓓ Ⓔ	17 Ⓐ Ⓑ Ⓒ Ⓓ Ⓔ	27 Ⓐ Ⓑ Ⓒ Ⓓ Ⓔ
8	Ⓐ Ⓑ Ⓒ Ⓓ Ⓔ	18 Ⓐ Ⓑ Ⓒ Ⓓ Ⓔ	28 Ⓐ Ⓑ Ⓒ Ⓓ Ⓔ
9	Ⓐ Ⓑ Ⓒ Ⓓ Ⓔ	19 Ⓐ Ⓑ Ⓒ Ⓓ Ⓔ	29 Ⓐ Ⓑ Ⓒ Ⓓ Ⓔ
10	Ⓐ Ⓑ Ⓒ Ⓓ Ⓔ	20 Ⓐ Ⓑ Ⓒ Ⓓ Ⓔ	30 Ⓐ Ⓑ Ⓒ Ⓓ Ⓔ

ASVAB Practice Test 2
Mathematics Knowledge

1	Ⓐ Ⓑ Ⓒ Ⓓ Ⓔ	11 Ⓐ Ⓑ Ⓒ Ⓓ Ⓔ	21 Ⓐ Ⓑ Ⓒ Ⓓ Ⓔ
2	Ⓐ Ⓑ Ⓒ Ⓓ Ⓔ	12 Ⓐ Ⓑ Ⓒ Ⓓ Ⓔ	22 Ⓐ Ⓑ Ⓒ Ⓓ Ⓔ
3	Ⓐ Ⓑ Ⓒ Ⓓ Ⓔ	13 Ⓐ Ⓑ Ⓒ Ⓓ Ⓔ	23 Ⓐ Ⓑ Ⓒ Ⓓ Ⓔ
4	Ⓐ Ⓑ Ⓒ Ⓓ Ⓔ	14 Ⓐ Ⓑ Ⓒ Ⓓ Ⓔ	24 Ⓐ Ⓑ Ⓒ Ⓓ Ⓔ
5	Ⓐ Ⓑ Ⓒ Ⓓ Ⓔ	15 Ⓐ Ⓑ Ⓒ Ⓓ Ⓔ	25 Ⓐ Ⓑ Ⓒ Ⓓ Ⓔ
6	Ⓐ Ⓑ Ⓒ Ⓓ	16 Ⓐ Ⓑ Ⓒ Ⓓ Ⓔ	
7	Ⓐ Ⓑ Ⓒ Ⓓ	17 Ⓐ Ⓑ Ⓒ Ⓓ Ⓔ	
8	Ⓐ Ⓑ Ⓒ Ⓓ	18 Ⓐ Ⓑ Ⓒ Ⓓ	
9	Ⓐ Ⓑ Ⓒ Ⓓ Ⓔ	19 Ⓐ Ⓑ Ⓒ Ⓓ	
10	Ⓐ Ⓑ Ⓒ Ⓓ Ⓔ	20 Ⓐ Ⓑ Ⓒ Ⓓ Ⓔ	

ASVAB Math Practice Test 1

CAT-ASVAB

Arithmetic Reasoning

- **16 questions**
- **Total time for this section:** 39 Minutes
- **Calculators are not allowed for this test.**

1) Aria was hired to teach three identical math courses, which entailed being present in the classroom 36 hours altogether. At $25 per class hour, how much did Aria earn for teaching one course?

   A. $50

   B. $300

   C. $600

   D. $1,400

2) Karen is 9 years older than her sister Michelle, and Michelle is 4 years younger than her brother David. If the sum of their ages is 82, how old is Michelle?

   A. 21

   B. 25

   C. 29

   D. 23

3) John is driving to visit his mother, who lives 300 miles away. How long will the drive be, round-trip, if John drives at an average speed of 50 mph?

   A. 95 Minutes

   B. 260 Minutes

   C. 645 Minutes

   D. 720 Minutes

4) Julie gives 8 pieces of candy to each of her friends. If Julie gives all her candy away, which amount of candy could have been the amount she distributed?

   A. 187

   B. 216

   C. 343

   D. 223

5) If a rectangle is 30 feet by 45 feet, what is its area?

   A. 1,350

   B. 1,250

   C. 1,000

   D. 750

6) You are asked to chart the temperature during an 8 hour period to give the average. These are your results:

   7 am: 2 degrees   11 am: 32 degrees

   8 am: 5 degrees   12 pm: 35 degrees

   9 am: 22 degrees   1 pm: 35 degrees

   10 am: 28 degrees   2 pm: 33 degrees

   What is the average temperature?

   A. 36   C. 24

   B. 28   D. 22

7) Each year, a cyber café charges its customers a base rate of $15, with an additional $0.20 per visit for the first 40 visits, and $0.10 for every visit after that. How much does the cyber café charge a customer for a year in which 60 visits are made?

   A. $25   C. $35

   B. $29   D. $39

8) If a vehicle is driven 32 miles on Monday, 35 miles on Tuesday, and 29 miles on Wednesday, what is the average number of miles driven each day?

   A. 32 Miles   C. 29 Miles

   B. 31 Miles   D. 27 Miles

9) Three co-workers contributed $10.25, $11.25, and $18.45 respectively to purchase a retirement gift for their boss. What is the maximum amount they can spend on a gift?

   A. $42.25   C. $39.95

   B. $40.17   D. $27.06

10) While at work, Emma checks her email once every 90 minutes. In 9-hour, how many times does she check her email?

   A. 4 Times

   B. 5 Times

   C. 7 Times

   D. 6 Times

11) A family owns 15 dozen of magazines. After donating 57 magazines to the public library, how many magazines are still with the family?

   A. 180

   B. 152

   C. 123

   D. 98

12) In the deck of cards, there are 4 spades, 3 hearts, 7 clubs, and 10 diamonds. What is the probability that William will pick out a spade?

   A. $\frac{1}{6}$

   B. $\frac{1}{8}$

   C. $\frac{1}{9}$

   D. $\frac{1}{5}$

13) What is the prime factorization of 560?

   A. $2 \times 2 \times 5 \times 7$

   B. $2 \times 2 \times 2 \times 2 \times 5 \times 7$

   C. 2×7

   D. $2 \times 2 \times 2 \times 5 \times 7$

14) William is driving a truck that can hold 5 tons maximum. He has a shipment of food weighing 32,000 pounds. How many trips will he need to make to deliver all of the food?

   A. 1 Trip

   B. 3 Trips

   C. 4 Trips

   D. 6 Trips

15) A man goes to a casino with $180. He loses $40 on blackjack, then loses another $50 on roulette. How much money does he have left?

    A. $75   C. $105

    B. $90   D. $120

16) A woman owns a dog walking business. If 3 workers can walk 9 dogs, how many dogs can 5 workers walk?

    A. 13   C. 15

    B. 14   D. 19

IF YOU FINISH BEFORE TIME IS CALLED, YOU MAY CHECK YOUR WORK ON THIS SECTION ONLY. DO NOT TURN TO OTHER SECTION IN THE TEST. **STOP**

ASVAB Math Practice Test 1

CAT-ASVAB

Mathematics Knowledge

- Total time for this section: 18 Minutes
- 16 questions
- Calculators are not allowed for this test.

1) If $a = 3$, what is the value of b in this equation?
$$b = \frac{a^2}{3} + 3$$

   A. 10
   B. 8
   C. 6
   D. 4

2) The eighth root of 256 is:

   A. 6
   B. 4
   C. 8
   D. 2

3) A circle has a radius of 5 inches. What is its approximate area? ($\pi = 3.14$)

   A. 90.7 square inches
   B. 78.5 square inches
   C. 31.4 square inches
   D. 25 square inches

4) If $-8a = 64$, then $a =$ ___

   A. -8
   B. 8
   C. 16
   D. 0

5) In the following diagram what is the value of x?

   A. 60°
   B. 90°
   C. 45°
   D. 15°

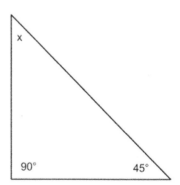

6) In the following right triangle, what is the value of x rounded to the nearest hundredth?

   A. 23.24

   B. 2.33

   C. 10.29

   D. 6.40

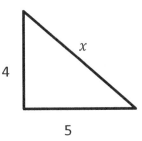

7) $(5x + 5)(2x + 6) = ?$

   A. $5x + 6$

   B. $10x^2 + 40x + 30$

   C. $5x + 5x + 30$

   D. $5x^2 + 5$

8) $5(a - 6) = 22$, what is the value of a?

   A. 2.4

   B. 10.4

   C. 7

   D. 11

9) If $3^{24} = 3^8 \times 3^x$, what is the value of x?

   A. 2

   B. 1.5

   C. 3

   D. 16

10) Which of the following is an obtuse angle?

   A. 116°

   B. 80°

   C. 68°

   D. 25°

11) Factor this expression: $x^2 + 5 - 6$

   A. $x2(5 + 6)$

   B. $x(x + 5 - 6)$

   C. $(x + 6)(x - 1)$

   D. $(x + 6)(x - 6)$

12) Find the slope of the line running through the points $(6, 7)$ and $(5, 3)$.

   A. $\frac{1}{4}$

   B. 4

   C. -4

   D. $-\frac{1}{4}$

13) What is the value of $\sqrt{100} \times \sqrt{36}$?

   A. 120

   B. $\sqrt{136}$

   C. 60

   D. $\sqrt{16}$

14) Which of the following is not equal to 5^2?

   A. the square of 5

   B. 5 squared

   C. 5 cubed

   D. 5 to the second power

15) The cube root of 2,197 is?

   A. 133

   B. 13

   C. 6.5

   D. 169

16) What is 952,710 in scientific notation?

   A. 95.271

   B. 9.5271×10^5

   C. 0.095271×10^6

   D. 0.95271

ASVAB Math Practice Test 2

Paper and Pencil-ASVAB

Arithmetic Reasoning

- Total time for this section: 36 Minutes
- 30 questions
- Calculators are not allowed for this test.

IF YOU FINISH BEFORE TIME IS CALLED, YOU MAY CHECK YOUR WORK ON THIS SECTION ONLY. DO NOT TURN TO ANY OTHER SECTION IN THE TEST. **STOP**

1) Will has been working on a report for 6 hours each day, 7 days a week for 2 weeks. How many minutes has Will worked on his report?

   A. 42

   B. 84

   C. 2,520

   D. 5,040

2) James is driving to visit his mother, who lives 340 miles away. How long will the drive be, round-trip, if James drives at an average speed of 50 mph?

   A. 135 minutes

   B. 310 minutes

   C. 741 minutes

   D. 816 minutes

3) In a classroom of 60 students, 42 are female. What percentage of the class is male?

   A. 34%

   B. 22%

   C. 30%

   D. 26%

4) You are asked to chart the temperature during a 6-hour period to give the average. These are your results:

   7 am: 7 degrees

   8 am: 9 degrees

   9 am: 22 degrees

   10 am: 28 degrees

   11 am: 28 degrees

   12 pm: 30 degrees

   What is the average temperature?

   A. 32.67

   B. 24.67

   C. 20.67

   D. 18.27

5) During the last week of track training, Emma achieves the following times in seconds: 66, 57, 54, 64, 57, and 59. Her three best times this week (least times) are averaged for her final score on the course. What is her final score?

   A. 56 seconds

   B. 57 seconds

   C. 59 seconds

   D. 61 seconds

6) How many square feet of tile is needed for a 15 $feet \times 15\ feet$ room?

   A. 225 square feet

   B. 118.5 square feet

   C. 112 square feet

   D. 60 square feet

7) With what number must 1.303572 be multiplied in order to obtain the number 1303.572?

   A. 100

   B. 1,000

   C. 10,000

   D. 100,000

8) Which of the following is NOT a factor of 50?

   A. 5

   B. 2

   C. 10

   D. 15

9) Emma is working in a hospital supply room and makes $25.00 an hour. The union negotiates a new contract giving each employee a 4% cost of living raise. What is Emma's new hourly rate?

   A. $26 an hour

   B. $28 an hour

   C. $30 an hour

   D. $31.50 an hour

10) Emily and Lucas have taken the same number of photos on their school trip. Emily has taken 4 times as many photos as Mia. Lucas has taken 21 more photos than Mia. How many photos has Mia taken?

   A. 7

   B. 9

   C. 11

   D. 13

11) Will has been working on a report for 5 hours each day, 6 days a week for 2 weeks. How many minutes has Will worked on his report?

   A. 7,444 minutes

   B. 5,524 minutes

   C. 3,600 minutes

   D. 2,640 minutes

12) Find the average of the following numbers: 22, 34, 16, 20

   A. 23

   B. 35

   C. 30

   D. 23.3

13) A mobile classroom is a rectangular block that is 90 feet by 30 feet in length and width respectively. If a student walks around the block once, how many yards does the student cover?

   A. 2,700 $yards$

   B. 240 $yards$

   C. 120 $yards$

   D. 60 $yards$

14) What is the distance in miles of a trip that takes 2.1 hours at an average speed of 16.2 miles per hour? (Round your answer to a whole number)

   A. 44 $miles$

   B. 34 $miles$

   C. 30 $miles$

   D. 18 $miles$

15) The sum of 6 numbers is greater than 120 and less than 180. Which of the following could be the average (arithmetic mean) of the numbers?

   A. 20

   B. 26

   C. 30

   D. 34

16) A barista averages making 15 coffees per hour. At this rate, how many hours will it take until she's made 1,500 coffees?

   A. 95 hours

   B. 90 hours

   C. 100 hours

   D. 105 hours

17) There are 120 rooms that need to be painted and only 12 painters available. If there are still 12 rooms unpainted by the end of the day, what is the average number of rooms that each painter has painted?

   A. 9

   B. 12

   C. 14

   D. 16

18) Nicole was making $7.50 per hour and got a raise to $7.75 per hour. What percentage increase was Nicole's raise?

   A. 2%

   B. 1.67%

   C. 3.33%

   D. 6.66%

19) An architect's floor plan uses $\frac{1}{2}$ inch to represent one mile. What is the actual distance represented by $4\frac{1}{2}$ inches?

   A. 9 *miles*

   B. 8 *miles*

   C. 7 *miles*

   D. 6 *miles*

20) A snack machine accepts only quarters. Candy bars cost 25¢, a package of peanuts costs 75¢, and a can of cola costs 50¢. How many quarters are needed to buy two Candy bars, one package of peanuts, and one can of cola?

   A. 8 quarters

   B. 7 quarters

   C. 6 quarters

   D. 5 quarters

21) The hour hand of a watch rotates 30 degrees every hour. How many complete rotations does the hour hand make in 8 days?

   A. 12

   B. 14

   C. 16

   D. 18

22) What is the product of the square root of 81 and the square root of 25?

   A. 2,025

   B. 15

   C. 25

   D. 45

23) If $2y + 4y + 2y = -24$, then what is the value of y?

   A. -3

   B. -2

   C. -1

   D. 0

24) A bread recipe calls for $2\frac{2}{3}$ cups of flour. If you only have $1\frac{5}{6}$ cups of flour, how much more flour is needed?

   A. 1

   B. $\frac{1}{2}$

   C. 2

   D. $\frac{5}{6}$

ASVAB Math Prep 2020-2021

25) Convert 0.023 to a percent.

   A. 0.2%

   B. 0.23%

   C. 2.30%

   D. 23%

26) Will has been working on a report for 3 hours each day, 7 days a week for 2 weeks. How many minutes has will worked on his report?

   A. 6,364 *minutes*

   B. 4,444 *minutes*

   C. 2,520 *minutes*

   D. 1560 *minutes*

27) A writer finishes 180 pages of his manuscript in 20 hours. How many pages is his average per hour?

   A. 18

   B. 6

   C. 3

   D. 9

28) Camille uses a 30% off coupon when buying a sweater that costs $50. If she also pays 5% sales tax on the purchase, how much does she pay?

   A. $35

   B. $36.75

   C. $39.95

   D. $47.17

29) I've got 34 quarts of milk and my family drinks 2 gallons of milk per week. How many weeks will that last us?

   A. 2 *weeks*

   B. 2.5 *weeks*

   C. 3.25 *weeks*

   D. 4.25 *weeks*

30) A floppy disk shows 937,036 bytes free and 739,352 bytes used. If you delete a file of size 652,159 bytes and create a new file of size 599,986 bytes, how many free bytes will the floppy disk have?

A. 687,179

B. 791,525

C. 884,867

D. 989,209

IF YOU FINISH BEFORE TIME IS CALLED, YOU MAY CHECK YOUR WORK ON THIS SECTION ONLY. DO NOT TURN TO OTHER SECTION IN THE TEST. **STOP**

ASVAB Math Practice Test 2

Paper and Pencil-ASVAB

Mathematics Knowledge

- 25 questions
- **Total time for this section:** 24 Minutes
- **Calculators are not allowed for this test.**

1) $(x+7)(x+5) = ?$

   A. $x^2 + 12x + 12$

   B. $2x + 12x + 12$

   C. $x^2 + 35x + 12$

   D. $x^2 + 12x + 35$

2) Convert 670,000 to scientific notation.

   A. 6.70×1000

   B. 6.70×10^{-5}

   C. 6.70×100

   D. 6.7×10^5

3) What is the perimeter of the triangle in the provided diagram?

   A. 15,625

   B. 625

   C. 75

   D. 25

   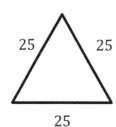

4) If x is a positive integer divisible by 6, and $x < 60$, what is the greatest possible value of x?

   A. 54

   B. 48

   C. 36

   D. 59

5) There are two pizza ovens in a restaurant. Oven 1 burns four times as many pizzas as oven 2. If the restaurant had a total of 15 burnt pizzas on Saturday, how many pizzas did oven 2 burn?

   A. 3

   B. 6

   C. 9

   D. 12

6) Which of the following is an obtuse angle?

   A. 56°

   B. 72°

   C. 123°

   D. 211°

7) $7^7 \times 7^8 = ?$

   A. 7^{56}

   B. $7^{0.89}$

   C. 7^{15} 15

   D. 1^7

8) What is 5231.48245 rounded to the nearest tenth?

   A. 5231.482

   B. 5231.5

   C. 5231

   D. 5231.48

9) The cube root of 512 is?

   A. 8

   B. 88

   C. 888

   D. 134,217,728

10) A circle has a diameter of 16 inches. What is its approximate area? ($\pi = 3.14$)

    A. 200.96

    B. 100.48

    C. 64.00

    D. 12.56

11) Which of the following is the correct calculation for 7!?

    A. $7 \times 6 \times 5 \times 4 \times 3 \times 2 \times 1$

    B. $1 \times 2 \times 3 \times 4 \times 5 \times 6$

    C. $0 \times 1 \times 2 \times 3 \times 4 \times 5 \times 6 \times 7$

    D. $1 \times 2 \times 3 \times 4 \times 5 \times 6 \times 7 \times 8$

12) The equation of a line is given as: $y = 5x - 3$. Which of the following points does not lie on the line?

   A. $(1, 2)$ C. $(3, 18)$

   B. $(-2, -13)$ D. $(2, 7)$

13) How long is the line segment shown on the number line below?

   A. -9 C. 8

   B. -8 D. 9

14) What is the distance between the points $(1, 3)$ and $(-2, 7)$?

   A. 3   C. 5

   B. 4   D. 6

15) $x^2 - 81 = 0$, x could be:

   A. 6   C. 12

   B. 9   D. 15

16) A rectangular plot of land is measured to be 160 feet by 200 feet. Its total area is:

   A. $32,000\ square\ feet$ C. $3,200\ square\ feet$

   B. $4,404\ square\ feet$ D. $2,040\ square\ feet$

17) With what number must 2.103119 be multiplied in order to obtain the number 21,031.19?

   A. 100

   B. 1,000

   C. 10,000

   D. 100,000

18) Which of the following is NOT a factor of 50?

   A. 5

   B. 10

   C. 2

   D. 100

19) The sum of 4 numbers is greater than 320 and less than 360. Which of the following could be the average (arithmetic mean) of the numbers?

   A. 80

   B. 85

   C. 90

   D. 95

20) One fourth the cube of 4 is:

   A. 25

   B. 16

   C. 32

   D. 8

21) What is the sum of the prime numbers in the following list of numbers?

   $$14, 12, 11, 16, 13, 20, 19, 36, 30$$

   A. 26

   B. 37

   C. 43

   D. 32

22) Convert 25% to a fraction.

   A. $\frac{1}{2}$

   B. $\frac{2}{3}$

   C. $\frac{1}{4}$

   D. $\frac{3}{4}$

23) The supplement angle of a 45° angle is:

  A. 135°
  B. 105°
  C. 90°
  D. 35°

24) 20% of 50 is:

  A. 30
  B. 25
  C. 20
  D. 10

25) Simplify: $5(2x^6)^3$.

  A. $10x^9$
  B. $10x^{18}$
  C. $40x^{18}$
  D. $40x^9$

IF YOU FINISH BEFORE TIME IS CALLED, YOU MAY CHECK YOUR WORK ON THIS SECTION ONLY. DO NOT TURN TO OTHER SECTION IN THE TEST. **STOP**

ASVAB Mathematics Practice Tests Answers and Explanations

ASVAB Practice Test 1 CAT - ASVAB			
Arithmetic Reasoning		Mathematics Knowledge	
1)	B	1)	C
2)	D	2)	D
3)	D	3)	B
4)	B	4)	A
5)	A	5)	C
6)	C	6)	D
7)	A	7)	B
8)	A	8)	B
9)	C	9)	D
10)	D	10)	A
11)	C	11)	C
12)	A	12)	B
13)	B	13)	C
14)	C	14)	C
15)	B	15)	B
16)	C	16)	B

ASVAB Math Practice Test 2 - Paper and Pencil

Arithmetic Reasoning				Mathematics Knowledge			
1)	D	16)	C	1)	D	16)	C
2)	D	17)	A	2)	D	17)	C
3)	C	18)	C	3)	C	18)	D
4)	C	19)	A	4)	A	19)	B
5)	A	20)	B	5)	A	20)	B
6)	A	21)	C	6)	C	21)	C
7)	B	22)	D	7)	C	22)	C
8)	D	23)	A	8)	B	23)	A
9)	A	24)	D	9)	A	24)	D
10)	A	25)	C	10)	A	25)	C
11)	C	26)	C	11)	A	26)	
12)	A	27)	D	12)	C	27)	
13)	B	28)	B	13)	D	28)	
14)	B	29)	D	14)	C	29)	
15)	B	30)	D	15)	B	30)	

ASVAB Math Practice Tests Explanations

In this section, answers and explanations are provided for two ASVAB Math Tests, Test I CAT-ASVAB and Test 2 Paper and Pencil – ASVAB. Review the answers and explanations to learn more about solving ASVAB Math questions fast.

ASVAB Math Practice Test 1
CAT-ASVAB Arithmetic Reasoning
Answers and Explanations

1) **Choice B is correct**

$36 \div 3 = 12$ hours for one course

$12 \times 25 = 300 \Rightarrow \$300$

2) **Choice D is correct**

$Michelle = Karen - 9, Michelle = David - 4, Karen + Michelle + David = 82$

$Karen + 9 = Michelle \Rightarrow Karen = Michelle - 9$

$Karen + Michelle + David = 82$

Now, replace the ages of Karen and David by Michelle. Then:

$Michelle + 9 + Michelle + Michelle + 4 = 82$

$3 Michelle + 13 = 82 \Rightarrow 3 Michelle = 82 - 13 \Rightarrow 3\, Michelle = 69 \Rightarrow Michelle = 23$

3) **Choice D is correct**

$distance = speed \times time \Rightarrow \text{time} = \dfrac{distance}{speed} = \dfrac{600}{50} = 12$

(Round trip means that the distance is 600 miles)

The round trip takes 12 hours. Change hours to minutes, then:

$$12 \times 60 = 720$$

4) Choice B is correct

Since Julie gives 8 pieces of candy to each of her friends, then, then number of pieces of candies must be divisible by 8.

A. $187 \div 8 = 23.375$
B. $216 \div 8 = 27$
C. $343 \div 8 = 42.875$
D. $223 \div 8 = 27.875$

Only choice b gives a whole number.

5) Choice A is correct

$Area\ of\ a\ rectangle = width \times length = 30 \times 45 = 1,350$

6) Choice C is correct

$average = \frac{sum}{total}$, $Sum = 2 + 5 + 22 + 28 + 32 + 35 + 35 + 33 = 192$

Total number of numbers = 8, $average = \frac{192}{8} = 24$

7) Choice A is correct

The base rate is $15. The fee for the first 40 visits is: $40 \times 0.20 = 8$

The fee for the visits 41 to 60 is: $20 \times 0.10 = 2$, Total charge: $15 + 8 + 2 = 25$

8) Choice A is correct

$average = \frac{sum}{total} = \frac{32+35+29}{3} = \frac{96}{3} = 32$

9) Choice C is correct

The amount they have = $\$10.25 + \$11.25 + \$18.45 = 39.95$

10) Choice D is correct

Change 9 hours to minutes, then: $9 \times 60 = 540\ minutes$

$\frac{540}{90} = 6$

11) Choice C is correct

15 dozen of magazines are 180 magazines: $15 \times 12 = 180$

$$180 - 57 = 123$$

12) Choice A is correct

$$probability = \frac{desired\ outcomes}{possible\ outcomes} = \frac{4}{4+3+7+10} = \frac{4}{24} = \frac{1}{6}$$

13) Choice B is correct

Find the value of each choice:

A. $2 \times 2 \times 5 \times 7 = 140$
B. $2 \times 2 \times 2 \times 2 \times 5 \times 7 = 560$
C. $2 \times 7 = 14$
D. $2 \times 2 \times 2 \times 5 \times 7 = 280$

14) Choice C is correct

$1\ ton = 2,000\ pounds,\quad 5\ ton = 10,000\ pounds,\ \frac{32,000}{10,000} = 3.2$

William needs to make at least 4 trips to deliver all of the food.

15) Choice B is correct

$180 - 40 - 50 = 90$

16) Choice C is correct

Each worker can walk 3 dogs: $9 \div 3 = 3$, 5 workers can walk 15 dogs. $5 \times 3 = 15$

ASVAB Math Practice Test 1
CAT-ASVAB Mathematics Knowledge
Answers and Explanations

1) Choice C is correct

If $a = 3$ then: $b = \frac{a^2}{3} + 3 \Rightarrow b = \frac{3^2}{3} + 3 = 3 + 3 = 6$

2) Choice D is correct

$\sqrt[8]{256} = 2,\quad (2^8 = 2 \times 2 \times 2 \times 2 \times 2 \times 2 \times 2 \times 2 = 256)$

3) Choice B is correct

(r = radius) Area of a circle $= \pi r^2 = \pi \times (5)^2 = 3.14 \times 25 = 78.5$

4) Choice A is correct

$-8a = 64 \quad \Rightarrow \quad a = \frac{64}{-8} = -8$

5) Choice C is correct

All angles in a triable add up to 180 degrees.

$90° + 45° = 135°$

$x = 180° - 135° = 45°$

6) Choice D is correct

Use Pythagorean Theorem: $a^2 + b^2 = c^2$

$(4)^2 + (5)^2 = c^2 \quad \Rightarrow \quad 16 + 25 = 41 = c^2 \Rightarrow C = \sqrt{41} = 6.403$

7) Choice B is correct

Use FOIL (first, out, in, last) method.

$(5x + 5)(2x + 6) = 10x^2 + 30x + 10x + 30 = 10x^2 + 40x + 30$

8) **Choice B is correct**

$$5(a-6) = 22 \Rightarrow 5a - 30 = 22 \Rightarrow 5a = 22 + 30 = 52$$
$$\Rightarrow 5a = 52 \Rightarrow a = \frac{52}{5} = 10.4$$

9) **Choice D is correct**

Use exponent multiplication rule: $x^a \cdot x^b = x^{a+b}$

Then: $3^{24} = 3^8 \times 3^x = 3^{8+x}$, $24 = 8 + x \Rightarrow x = 24 - 8 = 16$

10) **Choice A is correct**

An obtuse angle is an angle of greater than 90 degrees and less than 180 degrees. Only choice a is an obtuse angle.

11) **Choice C is correct**

To factor the expression $x^2 + 5 - 6$, we need to find two numbers whose sum is 5 and their product is -6. Those numbers are 6 and -1. Then: $x^2 + 5 - 6 = (x+6)(x-1)$

12) **Choice B is correct**

Slope of a line: $\frac{y_2 - y_1}{x_2 - x_1} = \frac{rise}{run}$, $\frac{y_2 - y_1}{x_2 - x_1} = \frac{3-7}{5-6} = \frac{-4}{-1} = 4$

13) **Choice C is correct**

$\sqrt{100} = 10$, $\sqrt{36} = 6$, $10 \times 6 = 60$

14) **Choice C is correct**

Only choice c is not equal to 5^2

15) **Choice B is correct**

$\sqrt[3]{2,197} = 13$

16) **Choice B is correct**

In scientific notation form, numbers are written with one whole number times 10 to the power of a whole number. Number 952,710 has 6 digits. Write the number and after the first digit put the decimal point. Then, multiply the number by 10 to the power of 5 (number of remaining digits). Then:

$$952,710 = 9.5271 \times 10^5$$

ASVAB Math Practice Test 2

Paper and Pencil - Arithmetic Reasoning

Answers and Explanations

1) Choice D is correct

2 $weeks$ = 14 $days$, then: $14 \times 6 = 84$ $hours$

$84 \times 60 = 5,040$ $minutes$

2) Choice D is correct

$distance = speed \times time \Rightarrow \text{time} = \frac{distance}{speed} = \frac{340+3}{50} = 13.6$

(Round trip means that the distance is 680 miles)

The round trip takes 13.6 hours. Change hours to minutes, then: $13.6 \times 60 = 816$

3) Choice C is correct

$60 - 42 = 18$ male students, $\quad \frac{18}{60} = 0.3$

Change 0.3 to percent $\Rightarrow 0.3 \times 100 = 30\%$

4) Choice C is correct

$average = \frac{sum}{total}$, Sum $= 7 + 9 + 22 + 28 + 28 + 30 = 124$

Total number of numbers $= 9$, $\frac{124}{6} = 20.67$

5) Choice A is correct

Emma's three best times are 54, 57, and 57. The average of these numbers is:

$average = \frac{sum}{total}$, Sum $= 54 + 57 + 57 = 168$, $\quad$ Total number of numbers $= 3$

$average = \frac{168}{3} = 56$

6) Choice A is correct

The area of a 15 $feet \times$ 15 $feet$ room is 225 square feet.

$$15 \times 15 = 225$$

7) Choice B is correct

$1.303572 \times 1000 = 1303.572$

8) Choice D is correct

The factors of 50 are: {1, 2, 5, 10, 25, 50}, 15 is not a factor of 50.

9) Choice A is correct

4 percent of 25 is: $25 \times \frac{4}{100} = 1$, Emma's new rate is $26.25 + 1 = 26$

10) Choice A is correct

$Emily = Lucas$, $Emily = 4\ Mia \Rightarrow Lucas = 4\ Mia$, $Lucas = Mia + 21$

then: $Lucas = Mia + 21 \Rightarrow 4\ Mia = Mia + 21$, Remove 1 Mia from both sides of the equation. Then: $3\ Mia = 21 \Rightarrow Mia = 7$

11) Choice C is correct

$12\ days, 12 \times 5 = 60\ hours, 60 \times 60 = 3{,}600\ minutes$

12) Choice A is correct

$Sum = 22 + 34 + 16 + 20 = 92$, $average = \frac{92}{4} = 23$

13) Choice B is correct

Perimeter of a rectangle $= 2 \times length + 2 \times width = 2 \times 90 + 2 \times 30 = 180 + 60 = 240$

14) Choice B is correct

$Speed = \frac{distance}{time}$, $16.2 = \frac{distance}{2.1} \Rightarrow distance = 16.2 \times 2.1 = 34.02$

Rounded to a whole number, the answer is 34.

15) Choice B is correct

Let's review the choices provided and find their sum.

a. $20 \times 6 = 120$
b. $26 \times 6 = 144 \Rightarrow$ is greater than 120 and less than 180
c. $30 \times 6 = 180$

d. $34 \times 6 = 204$

Only choice b gives a number that is greater than 120 and less than 180.

16) Choice C is correct

$$\frac{1\ hour}{15\ coffees} = \frac{x}{1500} \Rightarrow 15 \times x = 1 \times 1,500 \Rightarrow 15x = 1,500, \quad x = 100$$

It takes 100 hours until she's made 1,500 coffees.

17) Choice A is correct

$120 - 12 = 108, \frac{108}{12} = 9$

18) Choice C is correct

$$percent\ of\ change = \frac{chang}{original\ number}, \quad 7.75 - 7.50 = 0.25$$

$$percent\ of\ change = \frac{0.25}{7.50} = 0.0333 \Rightarrow 0.0333 \times 100 = 3.33\%$$

19) Choice A is correct

Write a proportion and solve.

$\frac{\frac{1}{2} inches}{4.5} = \frac{1\ mile}{x}$, Use cross multiplication, then: $\frac{1}{2}x = 4.5 \rightarrow x = 9$

20) Choice B is correct

Two candy bars costs 50¢ and a package of peanuts cost 75¢ and a can of cola costs 50¢. The total cost is: $50 + 75 + 50 = 175$, 175 is equal to 7 quarters. $7 \times 25 = 175$

21) Choice C is correct

Every day the hour hand of a watch makes 2 complete rotation. Thus, it makes 16 complete rotations in 8 days. $2 \times 8 = 16$

22) Choice D is correct

$\sqrt{81} \times \sqrt{25} = 9 \times 5 = 45$

23) Choice A is correct

$2y + 4y + 2y = -24 \Rightarrow 8y = -24 \Rightarrow y = -\frac{24}{8} \Rightarrow y = -3$

24) Choice D is correct

$$2\frac{2}{3} - 1\frac{5}{6} = 2\frac{4}{6} - 1\frac{5}{6} = \frac{16}{6} - \frac{11}{6} = \frac{5}{6}$$

25) Choice C is correct

To convert a decimal to percent, multiply it by 100 and then add percent sign (%).
$$0.023 \times 100 = 2.30\%$$

26) Choice C is correct

$2\ WEEKS = 14\ DAYS, 3\ hours \times 14\ Days = 42\ hours, 42\ hours = 2,520\ minutes$

27) Choice D is correct

$180 \div 20 = 9$

28) Choice B is correct

$$30\% \times 50 = \frac{30}{100} \times 50 = 15$$

The coupon has $15 value. Then, the selling price of the sweater is $35.50 - 15 = 35$

Add 5% tax, then: $\frac{5}{100} \times 35 = 1.75$ for tax, then: $35 + 1.75 = \$36.75$

29) Choice D is correct

$1\ quart = 0.25\ gallon, 34\ quarts = 34 \times 0.25 = 8.5\ gallons$, then: $\frac{8.5}{2} = 4.25$ weeks

30) Choice D is correct

The difference of the file added, and the file deleted is: $652,159 - 599,986 = 52,173$

$$937,036 + 52,173 = 989,209$$

ASVAB Math Practice Test 2

Paper and Pencil - Mathematics Knowledge

Answers and Explanations

1) **Choice D is correct**

   Use FOIL (First, Out, In, Last) method. $(x + 7)(x + 5) =$
   $x^2 + 5x + 7x + 35 = x^2 + 12x + 3$

2) **Choice D is correct**

   In scientific notation form, numbers are written with one whole number times 10 to the power of a whole number. Number 670,000 has 6 digits. Write the number and after the first digit put the decimal point. Then, multiply the number by 10 to the power of 5 (number of remaining digits). Then: $670,000 = 6.7 \times 10^5$

3) **Choice C is correct**

   Perimeter of a triangle $= side\ 1 + side\ 2 + side\ 3 = 25 + 25 + 25 = 75$

4) **Choice A is correct**

   From the choices provided, 36, 48 and 54 are divisible by 6. From these numbers, 54 is the biggest.

5) **Choice A is correct**

   $Oven\ 1 = 4\ oven\ 2,$ If Oven 2 burns 3 then oven 1 burns 12 pizzas. $3 + 12 = 15$

6) **Choice C is correct**

   An obtuse angle is an angle of greater than 90° and less than 180°.

7) **Choice C is correct**

   Use exponent multiplication rule: $x^a \cdot x^b = x^{a+b}$

   Then: $7^7 \times 7^8 = 7^{15}$

8) Choice B is correct

5231.48245 rounded to the nearest tenth equals 5231.5

(Because 5231.48 is closer to 5,231.5 than 5,231.4)

9) Choice A is correct

$$\sqrt[3]{512} = 8$$

10) Choice A is correct

$Diameter = 16$, then: $Radius = 8$

Area of a circle $= \pi r^2 \Rightarrow A = 3.14(8)^2 = 200.96$

11) Choice A is correct

$7! = 7 \times 6 \times 5 \times 4 \times 3 \times 2 \times 1$

12) Choice C is correct

Let's review the choices provided. Put the values of x and y in the equation.

A. $(1,2) \Rightarrow x = 1 \Rightarrow y = 2$ This is true!

B. $(-2,-13) \Rightarrow x = -2 \Rightarrow y = -13$ This is true!

C. $(3,18) \Rightarrow x = 3 \Rightarrow y = 12$ This is not true!

D. $(2,7) \Rightarrow x = 2 \Rightarrow y = 7$ This is true!

13) Choice D is correct

$1 - (-8) = 1 + 8 = 9$

14) Choice C is correct

Use distance formula:

$d = \sqrt{(x_1 - x_2)^2 + (y_1 - y_2)^2} = \sqrt{(1-(-2))^2 + (3-7)^2}$

$\sqrt{9+16} = \sqrt{25} = 5$

15) Choice B is correct

$x^2 2 - 81 = 0 \Rightarrow x^2 = 81 \Rightarrow x$ could be 9 or -9.

16) Choice C is correct

$Area\ of\ a\ rectangle = width \times length = 160 \times 200 = 3,200$

17) Choice C is correct

Number 2.103119 should be multiplied by 10,000 in order to obtain the number 21,031.19

$$2.103119 \times 10,000 = 21,031.19$$

18) Choice D is correct

factor of 50 = {1, 2, 5, 10, 25, 50}

100 is not a factor of 50.

19) Choice B is correct

Let's review the choices provided.

A. $80 \times 4 = 320$
B. $85 \times 4 = 340$
C. $90 \times 4 = 360$
D. $95 \times 4 = 380$

From choices provided, only 340 is greater than 320 and less than 360.

20) Choice B is correct

The cube of $4 = 4 \times 4 \times 4 = 64$

$\frac{1}{4} \times 64 = 16$

21) Choice C is correct

From the list of numbers, 11, 13, and 19 are prime numbers. Their sum is:

$$11 + 13 + 19 = 43$$

22) Choice C is correct

$25\% = \frac{25}{100} = \frac{1}{4}$

23) Choice A is correct

Two Angles are supplementary when they add up to 180 degrees.

$135° + 45° = 180°$

24) Choice D is correct

$\frac{20}{100} \times 50 = 10$

25) Choice C is correct

$5(2x^6)^3 \Rightarrow 5 \times 2^3 \times x^{18} = 40x^{18}$

"Effortless Math Education" Publications

Effortless Math authors' team strives to prepare and publish the best quality ASVAB Mathematics learning resources to make learning Math easier for all. We hope that our publications help you learn Math in an effective way and prepare for the ASVAB test.

We all in Effortless Math wish you good luck and successful studies!

Effortless Math Authors

Visit www.EffortlessMath.com
for Online Math Practice

www.EffortlessMath.com

... So Much More Online!

✓ FREE Math lessons

✓ More Math learning books!

✓ Mathematics Worksheets

✓ Online Math Tutors

Need a PDF version of this book?

Visit www.EffortlessMath.com